A FUTURE FOR
OUR COUNTRYSIDE

Co-writers of A Future for Our Countryside

Martin Elson
Philip Lowe
Derek Nicholls
Clive Potter
Bill Slee
Michael Winter

A FUTURE FOR OUR COUNTRYSIDE

John Blunden &
Nigel Curry

Basil Blackwell
In Association with the **Countryside** COMMISSION

Basil Blackwell Ltd
108 Cowley Road, Oxford, OX4 1JF, UK

Basil Blackwell Inc.
432 Park Avenue South, Suite 1503
New York, NY 10016, USA

British Library Cataloguing in Publication Data

Blunden, John
   A future for our countryside.
   1. England. Rural regions. Land use
   I. Title II. Curry, Nigel
   333.76´13´0942
   ISBN 0-631-16272-0

Library of Congress Cataloging in Publication Data

Blunden, John.
   A future for our countryside / John Blunden and Nigel Curry.
      p. cm.
   Bibliography: p.
   Includes index.
   ISBN 0-631-16272-0
   1. England – Rural conditions. 2. Wales – Rural conditions.
   3. Rural development – England. 4. Rural development – Wales.
   5. Regional planning – England. 6. Regional planning – Wales.
   I. Curry, Nigel. II. Title.
   HN398.E5B57 1988
   307.7´2´0942 – dc 19   88-19329
                                   CIP

Typeset in 10 on 11 pt Times
by Joshua Associates Limited, Oxford
Printed in Great Britain at the University Press, Cambridge

Contents

Acknowledgements

The editors would like to express their very warm thanks to the following people for their generous help and assistance in the realisation of this book:

Anna Dekker of the Countryside and Community Research Unit, Gloucestershire College of Art and Technology;
John Hunt, Project Officer in the Faculty of Social Sciences at the Open University;
Gillian Pritchard of the Photographic Library at the Countryside Commission;
Paul Smith, Liaison Librarian to the Faculty of Social Sciences in the Open University;
Dr Sadie Ward of the Museum of English Rural Life at the University of Reading.

We are also grateful to the many others who assisted in our search for pictures and to the secretaries in the Faculty of Social Sciences at the Open University who word processed the text of this book with speed and efficiency.

Contributors

JOHN BLUNDEN is Reader in Geography at the Open University. A graduate in Social Studies, his doctoral thesis investigated spatial and temporal variations in the impact of agricultural support policies on farm enterprise. After working as a television producer with the BBC, he returned to academic life through a fellowship at the University of Sussex. Since then his research interests, frequently sponsored by governments on both sides of the Atlantic, have covered a broad range of topics in the field of rural resource management and have been reflected in a considerable number of books and papers.

NIGEL CURRY is Reader in Environmental Studies at Gloucestershire College of Arts and Technology and Research Associate at the Royal Agricultural College, Cirencester. He has degrees in economics and agricultural economics and holds a Ph.D. in land economy from the University of Cambridge. His previous books include *The Changing Countryside* and *The Countryside Handbook* (with John Blunden). His current research interests are concerned with the implementation of recreation policies in the countryside and the analysis of environmental job markets in rural areas.

CLIVE POTTER is Lecturer in Environmental Management at Wye College in the University of London where he is currently researching the possible environmental effects of 'set-aside' policies in UK agriculture. He is European editor of the *International Yearbook of Rural Planning*. His previous publications include *Investing In Rural Harmony*, which proposed reforming the agricultural grant system to promote conservation and *The Countryside Tomorrow* a manifesto for the future of the countryside commissioned and published by the Royal Society for Nature Conservation.

MARTIN ELSON is a Professor in the School of Planning at Oxford Polytechnic where he teaches in the areas of environmental and countryside planning. The author of a national study of Green Belts published in 1986, his current interests include the development of policies to manage the impact of noisy sports in the countryside and work on the scope for farm diversification in the face of restrictive countryside policies. He is also Director of the newly-formed Oxford Centre for Tourism and Leisure Studies at Oxford Polytechnic.

DEREK NICHOLLS is Lecturer in Land Economy at the University of Cambridge and Fellow and Senior Tutor at Wolfson College, Cambridge. After graduating from Cambridge in rural estate management he undertook doctoral research into aspects of private forestry policy. Before taking up his present appointment he spent several years at the University of Glasgow, primarily lecturing in town and regional planning. His main research interests include forestry and other aspects of rural policy, land policy and taxation, and the land development process.

PHILIP LOWE is Lecturer in Countryside Planning at University College, London. He has degrees in Natural Science, Science Policy and History and has written extensively in the fields of land use planning, rural sociology, environmental politics and the history of ecology. A contributor to *The Changing Countryside*, his latest book (with G. Cox and M. Winter) is entitled *The Voluntary Principle in Conservation – a Study of the Farming and Wildlife Advisory Group* and is to be published late in 1988.

BILL SLEE is Senior Lecturer in Rural Resource Management at Seale-Hayne College which is a constituent part of Plymouth Polytechnic's Department of Agriculture. A graduate of the University of Cambridge, he moved to Aberdeen University to undertake work in the field of agricultural economics for which he obtained his doctorate. His current research interests include rural policy and farm diversification. He was one of the contributing authors of *The Changing Countryside* and has recently had published a book entitled *Alternative Farm Enterprises*.

MICHAEL WINTER is Director of the Centre for Rural Studies at the Royal Agricultural College, Cirencester. He has a degree in Rural Environment Studies from Wye College in the University of London. His doctoral thesis for the Open University explored the sociology of agricultural change in West Devon. Subsequently he has conducted research on many aspects of agricultural policy and nature conservation. He has also collaborated with Philip Lowe in the preparation of the volume on FWAG to be published at the end of 1988.

Preface

A new policy era

When *The Changing Countryside* appeared in 1985, as editors we were able to set out its rationale clearly and simply. We wrote of a countryside which had evolved through centuries of change which had reflected the interplay between the society of the time and its technology. Yet we recognized that by the early 1980s, change was occurring more rapidly and with greater consequences for the people of the country and the countryside of Britain than ever before. By analysing in depth four critical rural issues, those of agricultural expansion, the containment of settlements, the conservation of the wild and the welfare of our rural communities, we were able to look at the pros and cons of change and offer some alternative 'visions of the future' where we speculated about how change was likely to continue into the next decade.

Then in a final chapter we identified the changes that were most likely to occur in the next few years specifically in terms of possible public policy directions. We considered issues such as extending town and country planning controls over forestry and other primary land uses; how social planning and social opportunities in the countryside could be more effectively brought about; whether offering farmers compensation payments for not undertaking agricultural activities was really a cost-effective means of controlling rural change; and whether or not integrated rural development really could provide any more than a novelty as a new approach to rural adjustment.

Since the book was produced rural change has proceeded and people's expressions about a rural future have continued to appear at an unprecedented rate. The combined pressures of conservation groups and other rural lobbies, ably abetted by sympathetic media attention have begun to impact more directly on the policymakers who have been, in any case, much exercised about European Community (EC) overspending on agriculture. As we made clear in chapter 2 of *The Changing Countryside*, farm price support schemes which were absorbing more than 70 per cent of its annual budget in the early 1980s were resulting not only in radical changes in the appearance of the countryside, but producing major surpluses of dairy produce and cereals which could only be either stored or disposed of at knock-down prices to nations such as the USSR, much to the disenchantment of the general public.

A shift in rural policy emphasis by the government, particularly in the realms of agriculture, was given impetus at that time by a number of Tory backbenchers disassociating themselves from traditional farming and landowning interests. Richard Body, Conservative MP for Holland-with-Boston spearheaded this attack with a book published in 1982, called *Agriculture, the Triumph and the Shame*. Here, he undertook vehement criticism of the ranges of agricultural support policies and made the case for a more monetarist free trade approach to agriculture. From this basis, agricultural subsidies became a target for right-wing MPs, the monetarists in the Cabinet, and the Treasury. The Prime Minister herself was credited with the belief that 'an attack on farmers privileges is long overdue' (*Sunday Times*, 19 February, 1984).

As the power of farmers began to decline in the second Conservative administration, the conservation lobby began to increase its influence. In 1979 Mrs Thatcher had sought to develop strategies that would 'reduce over-sensitivity to environmental considerations' (*Sunday Times*, 18 November 1979), but by 1984 she was revising her opinions about the political significance of environmental concern, particularly in the light of the electoral successes of the Green Parties in Western Europe. Bad publicity over acid rain and nuclear waste together with accusations of environmental complacency by the Royal Commission on Environmental Pollution, had given impetus to these revisions – 'Tories seek to win environmentalist vote' ran the *Financial Times* (16 July 1984), after a series of ministerial meetings examining Britain's environmental policy. Following on closely from this, both the Ministry of Agriculture, Fisheries and Food (MAFF) and the Department of the Environment (DoE) set up environmental co-ordination units and a countryside research officer was appointed by Conservative Central Office.

In the sphere of land-use policy, Conservative backbenchers had begun to express conservation concerns a year earlier when government circulars to liberalize development around towns were drafted. These 1984 Circulars, *Green Belts* (14/84) and *Land for Housing*, (15/84) were much watered down when more than 60 Tory MPs, mainly from the Shire counties, registered their opposition to elements in the first drafts in support of a campaign in favour of retaining strict land-use controls in rural areas, orchestrated by the Council for the Protection of Rural England. We return to these Circulars in the form that they were approved, in chapter 1 and in chapter 4 of this book.

Back on the environmental front, the conservation theme in government circles was gaining further prominence. By July 1984, 166 MPs, mainly Conservatives, had signed an early-day motion put down by Conservative MP Andrew Hunter, again not without the influence of the Council for the Protection of Rural England, calling on the government 'to ensure that agricultural policy and the structure of public funding is widened so as to take full account of the need to protect and enhance the environment'.

Similar views appeared in two pamphlets, *Conserving the Countryside: A Tory View*, published by Conservative Central Office in May 1984 and written by Kenneth Carlisle, a Suffolk farmer and MP for Lincoln; and *Conservation and the Conservatives*, published in October 1984 and written by Tony Paterson, the parliamentary liaison officer of the Conservative Bow Group. These detailed the continuing losses of wildlife and habitats and called for the balance to be restored between agriculture and conservation, including a major overhaul of farming policies and the strengthening of safeguards in the 1981 Wildlife and Countryside Act. The Conservative Party, it seemed, was re-adjusting its historic attitude towards the land in a way that balanced the old Tory paternalism and the new monetarism: preparing on the one hand to give greater priority to countryside conservation and on the other, to subject agriculture to the rigours of greater competition and less state aid.

One of the first policy changes which offered some small adjustment in this balance between agriculture and conservation appeared in relation to the Broadlands area of Norfolk. Such was the rate of loss of the grazing marshlands, a key feature of the traditional landscape of the Broads, merely to produce cereal crops already in surplus, that in 1985 the government agreed to an experimental scheme run by the Countryside Commission and MAFF whereby farmers would be paid a sum per hectare to graze cattle by traditional means and not to convert marshland pasture to arable.

The success of this initially short-term scheme led to the development of the concept of Environmentally Sensitive Areas where in any one of a number of designated tracts of countryside of outstanding quality, grants are available to farmers to maintain the traditional landscape. We discuss these in detail in chapter 7.

Another significant illustration of changing attitudes of the government to the countryside has been the recent appearance of two significant policy statements. First there was the Agriculture Act of 1986. This charges agriculture ministers with new duties. Under Section 17, as well as promoting a stable and efficient agriculture industry, they are to develop the social and economic interests of rural areas, pay regard to conservation and positively promote the enjoyment of the countryside. We look at these new duties in more detail in chapter 1 and cast a critical eye over their implications in chapter 5.

Second, there was the introduction of the 1987 Government Circular, *Development Involving Agricultural Land* (16/87). In its draft form this for the first time laid before Parliament policy proposals that provide for significant moves away from agriculture as the dominant rural land use. The sacrosanct nature of agricultural land, which we shall see in chapter 1 has been at the centre of all governments' rural policies since the Second World War, was now to be relinquished in favour of wider economic and environmental goals.

If this Circular was particularly motivated by the increasing embarassment of agricultural food surpluses and the cost of the Common Agricultural Policy, it did not however, propose reforms to the Common Agricultural Policy itself, since this is the responsibility, chiefly, of the European Commission rather than any one member country. For a wide variety of reasons, most of them political, European Community policy change is grinding more slowly than the pace of policy change in Britain.

Now that some of our earlier ideas from *The Changing Countryside* about policy and land-use change in the countryside are becoming formally considered by the political establishment, there is, as you might expect, some competition, and even confusion, as to who should take the lead in initiating policy change within Westminster. The obvious importance of the 1987 Circular *Development Involving Agricultural Land* (16/87) and its repercussions, led to just such a situation.

But this was not conventional inter-party rivalry. Rather, it was the rivalry of two Ministers within the same government. Following a cabinet decision in March 1986, an inter-departmental working group (Alternative Land Use and the Rural Economy – ALURE) was set up. Its behind-closed-doors deliberations were dominated by officials of MAFF keen to propose alternative ways of keeping up farmers' incomes. DoE Ministers and senior officials remained, until late in the year, unaware of the significance of the wider debate going on. In December, ministers were denying the existence of ALURE, while national newspapers were, at the same time, 'leaking' its supposed conclusions. Relations between MAFF and the DoE remained poor.

The Circular *Development Involving Agricultural Land* was produced in its draft form by the DoE under the signature of Nicholas Ridley, the then Secretary of State for the Environment. Before it reached the House of Commons, however, Michael Jopling, the then Minister of Agriculture, announced new draft policies that went beyond the immediate spirit of the draft Circular in proposing wide-ranging changes to the rural economy, away from the maximization of agricultural production.

Junior DoE ministers accused MAFF of 'jumping the gun'. Newspapers described a 'land bungle' with ministers lambasted by Mrs Thatcher for their presentational shortcomings. Clearly, if any rural change was to move away from agriculture, it would be of primary importance to the DoE particularly since an important motivation for the draft circular was to consider ways in which potentially surplus farm land could be put to other uses.

After leaks about fresh policy initiatives in the press and challenges about them in the House of Commons, however, a number of policy documents were produced in March 1987 in a single folder, entitled *Farming and Rural Enterprise*. Since it had originated from the interdepartmental working group it is commonly known as the 'ALURE package'. These documents were to provide more details of the policy implications of the draft Circular, and we outline their main points in chapter 1. The folder as a whole now represents a consensus government view of a large number of policies relating to the countryside and was signed by no fewer than five ministers. The documents together call, for example, for the re-use of existing buildings in the countryside to create more jobs, the liberalizing of planning controls over lower grades of farmland, the development of new space-extensive activities in the countryside such as golf courses, a slackening of MAFF involvement in the determination of planning applications, and so on. These policy documents no doubt contributed to the approval of the Circular *Development Involving Agricultural Land* (16/87) in May 1987.

The production of these documents has required the two Ministries of Agriculture and Environment to work together – something which itself is quite new and clearly not without its problems. Thus, in the new Thatcher administration of June 1987 there is a Minister of State for Agriculture, John Gummer, who is responsible for the countryside, rural diversification, the environment and conservation. There is also, though, a Minister of State for the Environment, Lord Belstead, who had previously been a junior minister in MAFF, with a responsibility for the countryside, the environment and conservation. These posts represent a substantial shift in the organization of the government in response to significant changes in the structure of the countryside. Clearly the holders of these two posts will have to work together in the development of new rural policies.

Policies in profusion

Many of the speculations that were made in *The Changing Countryside* are therefore now being given explicit attention by the government, and indeed in some instances are actually coming to pass, driven by both the 1986 Agriculture Act and the policy documentation of *Farming and Rural Enterprise*. Let us briefly sample, then, a flavour of these government policies and proposals, together with other recent statements by the government and other agencies, concerning the countryside. These will be assessed in some detail in chapter 1 and critically evaluated in other parts of the book.

Two significant documents in *Farming and Rural Enterprise* are the DoE's review of policies for rural areas entitled *Rural Enterprise and Development* and the MAFF's *Farming U.K.*. These documents stress the need for the countryside now to be considered as a more broadly-based economy and environment. Both explain the background to the 1987 Circular *Development Involving Agricultural Land* (16/87) and also draw together some earlier Circulars on housing, industrial development and the re-use of buildings.

In the year prior to these policy statements, in addition to the government's passing of the 1986 Agriculture Act, the DoE also produced a comprehensive review of the government's conservation policies in a document entitled *Conservation and Development, the British Approach*. This was an outline of how Britain was responding to the principles laid out in the *World Conservation Strategy*, itself produced in 1980, and summarized conservation developments in agriculture, forestry, land use and development.

As we have already suggested, outside of the government itself, other agencies were not silent in making their proposals for rural readjustment. In 1984, the Chairman of the Countryside Commission for England and Wales[1] called for a White Paper on the 'Rural Estate'. Although this was turned down by a DoE minister, William Waldegrave, possibly because of the pending developments in the *Farming and Rural Enterprise* package and the setting up of ALURE, the Commission nevertheless set up its own expert group, the Countryside Policy Review Panel. It well understood 'the compelling need to examine most carefully the rapidly changing rural scene in England and Wales'. The report of the panel, *New Opportunities for the Countryside* was produced in May 1987 and contains proposals for wide-ranging changes to agriculture, forestry and woodland, recreation, urban pressure, and conservation. At the same time, changes in countryside recreation policy were being given more specific attention by the Countryside Commission in their recreation policy review, *Recreation 2000*.

The Agriculture Committee of the National Economic Development Office was concurrently charged by government to assess the implications of new and changing land uses in the countryside as a result of changing agricultural policies, particularly those concerned to take land out of agriculture. Their conclusions were produced in June 1987 under the title, *Directions for Change: Land Use in the 1990s*. In the report, the Committee recognizes that agricultural land surplus is likely to arise as a result of processes of change and it examines the potential of a number of rural land-use options, from alternative agricultural uses to industrial, forestry and leisure uses.

To complement this stage of policy thinking by the government and its agencies there have also been a number of official documents introducing ideas and schemes concerning how rural policy change might be implemented, a theme that we will return to in chapter 8. These include the *Farm Countryside Initiative* produced jointly by MAFF, the Department of Employment, the Manpower Services Commission and the DoE and a number of quangos, which outlines how the Manpower Services Commission's Community Programme can be developed in rural areas. By February 1988, however, the Minister for Employment had abandoned the Community Programme and in the light of this, the *Farm Countryside Initiative* will have to be modified.

The Development Commission, the main agency for promoting social and economic development in rural England, also produced a document in 1987, *Action for Rural Enterprise*. This outlines the latest government schemes available for industrial development in the countryside, and was one of the documents in the government's 1987 policy folder, *Farming and Rural Enterprise*. In April 1987, the MAFF introduced its first consultation paper

1 Whenever we refer to the Countryside Commission in this volume, we are talking about the Countryside Commission for England and Wales. We will mention the Countryside Commission for Scotland specifically by name when we refer to it.

concerning the reformulation of structural support for farms, entitled *The Farm Diversification Scheme*, clearly intended to encourage farmers away from producing foods currently in surplus. Indicative of these changes away from agricultural production, in 1987 MAFF combined and renamed two branches of its Agricultural Development Advisory Service, (the principal implementation advisory body for farmers). The Land and Water Service and the Agriculture Service were given the new title the Farm and Countryside Service.

Finally, at a European level, some agricultural reform policies are beginning to filter through to the domestic economy. These will also undoubtedly have a significant impact on rural areas. The Community Extensification Scheme, for example, is to encourage farmers to move away from surplus food production and set land aside from agricultural use. A new Socio-economic Structures Directive also recently has been announced by John McGregor, the post-June 1987 Minister of Agriculture, that seeks to implement, from Europe, further structural reforms in agriculture, away from the maximization of food production.

Since the production of *The Changing Countryside*, our speculations about the distant future have become the immediate concerns of government. This now allows us to look more closely and with greater definition at what is likely to happen to and in our countryside, to the turn of the century.

The structure of the book

In essence, this book, *A Future for our Countryside*, therefore takes up where *The Changing Countryside* left off. Firstly, in chapter 1 and the final chapter we examine in some detail exactly what the public policy response to a changing countryside has been since 1985. Within this theme, we examine policy ideas, concepts and compatibilities as well as, in the final chapter, making assessments of their impact and examining exactly how they might be implemented and resourced in practice. Chapter 1 is essentially a descriptive review of policy change, that is built upon by the critical appraisals of specific areas in chapters 2–7. Rather than simply reviewing a number of policy documents, in chapter 1 we adopt a thematic approach, examining how various policies influence particular aspects of rural life. This allows these themes to be developed in the remainder of the volume.

In chapters 2–7 we focus more closely on a number of initiatives and activities that have now come to the forefront of rural policy thinking. These include, in chapter 2, the specific consideration of measures that are currently being introduced to help restrict agricultural over-production, and how effective they are likely to be. The notion of agricultural 'set-aside' (that is, land required to be taken out of agricultural use), and its consequences is given particular attention in this context.

Forestry and woodland have become more important as alternative land uses, and the role they might have to play is assessed in chapter 3, both in terms of economic and environmental impacts. In chapter 4, we move on to consider how the planning of development and land uses is being modified, and consider whether a relaxation of development controls is likely to have detrimental conservation consequences at the same time as stimulating the rural economy.

The rural economy itself is an issue which we address directly in chapter 5, where we look specifically at a number of potentials and limitations associated with diversifying the rural economy. Recreation and tourism are evaluated in chapter 6 in terms of their likely benefits to both the rural economy and the participant, and in chapter 7, a number of conservation themes are drawn together and evaluated against a backcloth of changing political attitudes towards the countryside.

The final chapter has two main components. All of the policies and proposals evaluated in the body of the book, if they are to be meaningful, have to be resourced and implemented in some way. The final chapter therefore assesses how this might be achieved in practice. And in returning to the first policy appraisal theme, evaluates new initiatives and changing policy emphases discussed in the book, and considers how they fit together in an overall picture.

Will the sum of these changes lead to environmental and economic improvements in the countryside, or will one be at the expense of the other? What are the political consequences of such changes? Have there been any major omissions on the policy agenda? Does the administration of rural affairs require readjustment before such policies can operate effectively? What will be the impact of these policies on social welfare for people living and working in the countryside, particularly the socially and economically disadvantaged? Will they work to counter rural deprivation and poverty? Will policies tend to favour affluent lowland Britain rather than remoter rural areas?

In short, we are seeking to determine the environmental, economic and social characteristics of government policies and proposals that will determine *A Future for our Countryside*.

Chapter 1

From the changing countryside to a changing policy response

The changing countryside

New policies and proposals for the countryside put forward by government departments, quangos and other interested agencies and groups are appearing with increasing frequency. Many of them are beginning to offer ideas about how we can steer the countryside through a change that in the late 1980s has severally been termed a watershed in rural land-use policies, a period as significant as that of post-war reconstruction and, by the Countryside Commission's Policy Review Panel, a period that 'is more significant than at any time since the Enclosures'.

This chapter introduces the nature of the more important of these proposals, how they have come about, how they relate to each other and what their consequences are likely to be. Such a background is important for you, the reader, if you are to get the most out of our analyses of the new initiatives and issues critically evaluated in chapters 2 to 7. This chapter also lays the foundation for our evaluation of all of these policies and how they might be implemented, which we consider together at the end of the book.

Before we look at these policies and proposals, however, we start by very briefly reviewing some of the more significant changes in the countryside that have acted as a spur to this policy response and how such changes are leading to a breakdown of what we might have traditionally understood by such terms as 'rural' or 'the countryside'. If you have read *The Changing Countryside* both of these issues will provide a useful recap. In any event, they provide a useful starting point to an understanding of why such policies currently are being promulgated.

Central to all Government policies since the war has been the notion that all agricultural land is sacrosanct. The Scott report of 1942 provides the source of those much-quoted words 'every agricultural acre counts'. The development of the statutory town and country planning system that we have today, and the introduction of the economic planning systems for agriculture and forestry, both brought in immediately after the Second World War, were all predicated on the notion that agriculture and forestry should be given free reign to develop, and that by and large there should be a presumption against the development of buildings and settlements outside of the towns. This was perhaps understandable in the wake of a war that threatened our very

Figure 1.1 Ploughing the land for the war effort. During the Second World War every agricultural acre *did* count. But this was an attitude that was to persist for another forty years or more.
Source: MERL

existence through food shortages, and that used up over half of our nation's already-slender timber resources in the war effort.

This view, however, was still the prevailing one into the 1970s. The 1975 White Paper, *Food From Our Own Resources* reaffirmed the need to maintain and expand food production, and proposed that the retention of agricultural land in full production be made a special priority of the planning system. This was given official sanction in the 1976 Circular of the Department of the Environment (DoE), *Development Involving Agricultural Land* (75/76) the intention of which was severely to restrict the amount of land taken from agriculture for development.

In the mid-1970s, too, the Government set up an inter-departmental committee of senior civil servants to examine the problems and potentials of the countryside in a broader, more integrated way. The Countryside Review Committee in a number of topic reports, did examine conservation, recreation and rural community issues, but still returned to the inevitability, if not inviolability, of agricultural expansion as the salient priority for the countryside. Even as late as 1979 the White Paper *Farming and Nation* called for a further modest expansion in domestic food production. In the same year, Jerry Wiggin, the then Parliamentary Secretary to the Ministry of Agriculture, Fisheries and Food (MAFF) stated that 'the continuing loss of agricultural land is totally unacceptable'. In 1980 the Prime Minister, in an address at the Royal Agricultural Show, congratulated the farming industry on its ability to raise agricultural productivity.

Nevertheless, from the Second World War to the dawn of the 1980s these fundamentalist agricultural policies had always had their critics. The now-notorious minority report of the Scott Committee, by Professor Dennison in 1942, indicated that the preservation of agricultural land at all costs was likely to be sub-optimal in economic terms if it thwarted the opportunity for a more balanced and more populated rural economy. This might develop with a broader variety of industry and housing than that permitted only in the service of agriculture.

By the 1970s, *Food From Our Own Resources* had a wider range of critics. Here was a policy of agricultural protection and food expansion being proposed at the same time as food surpluses in Europe, especially within the European Community (EC) were already beginning to emerge. The folly of these types of policy, which we discussed at some length in *The Changing Countryside*, is now recognized by the establishment. It has been made clear not only by the problems of endemic, high-cost food surpluses, but also by other changes in important rural parameters.

The first of these has been demographic trends in the countryside. Although there have been continuing population losses in some of our remote areas, between 1961 and 1971 and between 1971 and 1981 the population of our rural areas overall grew very much faster than in our urban areas. National rural repopulation has been one spur to a rethinking of our fundamentalist agricultural policies.

Secondly, along with rural repopulation has come, despite restrictive planning policies, a broadening of the rural economic base. Employment in agriculture has been continually declining since the war, as a result of displacement by capital invested in labour-saving machinery and by the application of new technologies. So it is that agriculture now employs only 2.5 per cent of our working population. With more people living in rural areas pressure for non-agricultural jobs has grown with, in response, the development of small businesses, often of a high technology nature in the small towns and the deeper countryside.

The third factor leading to a substantial rethink of agricultural autonomy in rural areas, has been the pressure for development from urban areas. This was felt to be the largest threat of all to agricultural production after the war, but it has declined considerably into the 1980s. During the 1950s and 1960s some 15,000 hectares of farmland was taken for development each year, but

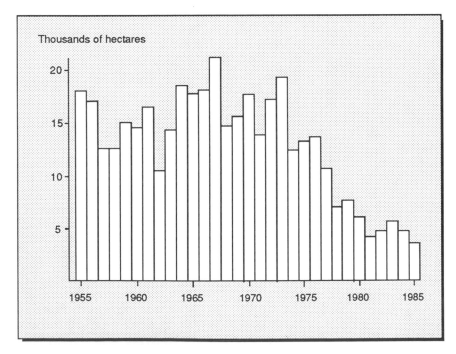

Figure 1.2 *Agricultural land loss 1955–85*. A steady decline in the amount of land going out of agriculture is one of the factors that has caused a substantial re-think in rural policies.
Source: R. H. Best; Department of the Environment

between 1980 and 1985 this had declined to less than 5,000 hectares per annum.

Aside from the endemic food surpluses in the EC, then, these three factors have been instrumental in causing a fundamental shift in policy thinking. The 'every agricultural acre counts' philosophy, prevailing until well into the 1980s has now been displaced by one of a strong economic base, more generally, for the countryside. As the DoE's *Rural Enterprise and Development* document states, 'The government believes that the best protection for the countryside is to ensure that the rural economy as a whole is strong and efficient'. These are almost exactly the words that Professor Dennison uttered in 1942.

So a number of important changes have taken place in the countryside that have caused a fundamental reconsideration of government policies. But does this mean that the countryside is now a significantly different place to what it was, say 10 years ago? It will be useful at the outset of this volume for us to spend a little time thinking about what we actually mean by the term 'the countryside' in the late 1980s, since this will have a significant bearing on the rest of the book. We do this next.

Defining 'the countryside'

The vast majority of books on the countryside have had a go at defining the terms 'rural' and 'countryside', and between them have managed to come up with a wide range of slightly different answers. The first thing that we can say about what we mean by the countryside in the late 1980s, then, is that there is no universally accepted definition of the term.

What we can do, however, is illustrate for you a range of these definitions, so that you can construct for yourself an understanding of the term 'countryside' – a term that undoubtedly can mean different things at different times for different people. We will also point to ways in which these terms 'rural' and 'countryside' are used variably in other parts of the book.

At its simplest, rural has been defined as anything that is left out once urban has been defined exhaustively. This has its limitations, however, since there are many parts of Britain that are not really either urban or rural, such as the urban fringe. Also, many would argue that urban and rural are not really opposites anyway, particularly if you think in sociological or cultural terms. You could certainly live on a farm and still have an urban mentality, for example.

Despite the problems associated with defining rural by 'playing it off' against definitions of urban, these types of definition are most common. Population censuses, for example distinguish urban and rural by reference to population densities. Some international definitions, for example that of the United Nations, distinguish urban and rural by defining the minimum size for a town.

A number of English structure plans have attempted to define degrees of 'rurality' by classifying places in terms of their distance from pre-specified urban centres. Also, definitions have included the analysis of the employment structure of an area – if it is dominated by rural types of employment, then it is considered to be 'countryside'. And all of these approaches are essentially spatial. If you choose to view the countryside from an anthropological perspective, a political one, a social one or even a vegetational one, a further range of definitions will result.

Moving now to slightly more complex definitions, the rural/urban continuum has been popular amongst a number of authors. This recognizes that rural and urban are not exclusive, but that any particular area is either more rural or more urban than another. Thus all areas fall on a continuum from very rural to very urban. Individual areas are assessed by a number of descriptive attributes of the social as well as the physical or spatial structure of the place, and these are then considered together, to come up with an overall picture.

Places are ordered on this continuum rather than given any absolute values. Such descriptions as 'unchanging', 'peasant', 'kinship', 'small communities', 'subsistence' and so on might be at the rural end of this continuum, but words such as 'commercial', 'cosmopolitan', 'large' and 'quickly-changing' might be at the urban end. You will probably realize that this process, whilst being more subtle than the numerical definitions suggested above, does not actually define rural at all, but simply tries to say how rural any one particular place might be.

Developing from this continuum model, a geographer from the University of Wales, Paul Cloke, has introduced a kind of 'statistical' continuum, known as the 'index of rurality'. Again this only assesses how rural one area is relative to another, but instead of using qualitative 'descriptors' he uses a number of statistical ones in combination. All of the statistical definitions we have discussed above, for example, are incorporated into this index, but it also includes population age structure, commuting patterns and household occupancy rates. Using all of these parameters, a continuum is developed by weighting them according to their rurality or urbanity, and then combining both values and weights together.

This approach to defining rurality has met with widest acceptance because of its comprehensiveness. Nevertheless, it still leads to rather restrictive definitions of rural. The Chiltern Hills are not considered rural, for example, using this index. Such an approach takes us little nearer, however, to defining rurality exhaustively since, like many other words in the English language, it means different things in different contexts. At least, though, you may be better equipped to recognize rurality when you see it!

Now that you are aware of some of the different types of parameters that contribute to a definition of rural, let us look at how we will be considering it in different contexts elsewhere in this book. Later in this chapter, we consider rurality in the context of the 'north-south divide' and suggest that there are at least two types of rural area in Britain, one of increasing growth and affluence and one that still suffers considerable deprivation and service loss.

In chapter 2, when we consider how farmers will fare in a climate of food over-production, we find it useful to consider three types of rural area. These are firstly, the agribusiness area, where farming will continue relatively unaffected by controls on production on our richest agricultural lands; secondly, our remote rural areas, where farming is likely to be protected by different types of public subsidy such as Less Favoured Area or Environmentally Sensitive Area grants; and thirdly, the 'middle band'. This is the area that is likely to fare worst in agricultural terms, since it is neither prosperous enough to weather the storms of cuts in production, nor poor enough to receive more direct forms of support.

In chapter 3, when we consider forests and woodlands, the countryside again takes on a number of guises. At one extreme we are concerned with vast tracts of upland countryside covered with commercial coniferous monoculture and at the other we are interested in the hedgerow trees and copses of the home counties so drastically damaged in the Autumn gales of 1987. In between we consider a range of ancient woodland that owes its geography to the luck of having survived the axe rather than any locational forethought.

When we come to talk in chapter 4 about the new pressures on the countryside brought about by the relaxation of planning controls, the countryside of critical interest now takes on the form of lowland England, particularly within striking distance of urban centres. This is where the pressures for development are greatest and the issues about new building in the countryside are at their most controversial. We even have a term, the *golden belt* or *golden horn* that defines a tract of countryside from Norfolk to Cornwall via London, where pressures for development through economic growth are particularly acute.

In considering how we might diversify the rural economy in chapter 5, we have to juggle our definition of rural to some extent, to fit in with existing employment and industrial statistics. These often do not distinguish between rural and other areas very well. Here, we do not want to think of rural in the narrowest way, since this might lead us to thinking about diversification only in terms of traditional rural industry. Most of the formal 'definitions' of rural that we have mentioned above are probably rather restrictive, when we come to think about diversification, so it will be important in reading chapter 5 to adopt a more liberal stance about the meaning of 'rural', than in other parts of the book.

In chapter 6, where we discuss recreation and leisure, the 'countryside' does connote an area outside of the urban boundary, since it is the contrast to the urban scene that commonly motivates countryside recreation participation. Apart from this qualification, any part of the land surface may be considered rural for the purposes of potential recreation use, but again we would have to limit our definition to those areas that allow some form of public access.

Finally, in chapter 7, the conservation policy debate seems firmly centred on a definition of rural that is concerned with areas of agricultural and environmental conflict. These are frequently specific pockets of land such as the Halvergate Marshes or a Site of Special Scientific Interest. In terms of the political impacts of legislation concerned with conservation, though, it would even be possible, at times, to define the Houses of Parliament as being steeped in rurality!

Having given you some idea of the different ways in which the term 'the countryside' can be considered, let us now turn to the sorts of policies and proposals that are being put forward by the government and others as a means of dealing with a number of discrete but related countryside changes.

Policies and proposals: curbing agricultural production

Food surpluses have certainly been the most powerful drive for rural policy reform. It has been recognized by the government in MAFF's 1987 *Farming U.K.* document that the principal cause of these surpluses has been support policies for agriculture. Support costs have increased tenfold in the last decade causing, through artificially high food prices, an imbalance between the supply and demand for food. At the same time, they have placed a severe strain on the EC budget as a whole since over 70 per cent of it has been spent on the agricultural sector.

But these support policies are not unique to Europe – most countries have policies that support agriculture in some way. This has actually created an over-production of food globally, although the ineffective distribution of it still devastatingly leads to famine in many parts of the world. In response to this world over-production, significant policy changes affecting Britain were agreed in September 1986 in a new round of multilateral trade negotiations under the General Agreement on Tariffs and Trade. These negotiations, agreed by all EC countries, now aim to correct distortions and instabilities in world markets, aim to achieve greater liberalization of trade and aim to increase international disciplines on all direct and indirect subsidies affecting trade. If adherence to these principles is not achieved, international trade wars in food are almost bound to ensue.

With these aims in mind, what is the British government now proposing as a means of curbing over-production and what measures more generally have been suggested by other bodies? To begin with, the government certainly does not intend to reduce advice, research and technical advances in the agricultural industry, despite the fact that many, including the Countryside Policy Review Panel, consider these to be a prime cause of over-production. It is, however, to shift the burden of cost of these types of development onto the industry more directly – a reflection of a developing Conservative 'free market' in agriculture.

Returning to the Ministry's *Farming U.K.*, a number of measures to curb over-production are discussed. Firstly, and perhaps most powerfully, it is clear that because it is the artificially high prices of food that are leading to surpluses, price restraint policies must be introduced. These are seen as long-term measures, because they can come about only whilst at the same time producing some support for farm incomes.

Their introduction will always be difficult since they are bound to have some adverse impact on farmers themselves. Nevertheless, the government is gearing up to this with successive Ministers of Agriculture that are increasingly 'dry' – from Peter Walker and Michael Jopling to John MacGregor, previously Chief Secretary to the Treasury and a Minister who strongly favours an adaptable farming industry, capable of standing on his own feet.

Such price restraint measures are considered to be potentially environmentally sensitive since they would lead to less intensive farming. The Countryside Policy Review Panel sees this as desirable, as long as the moves away from intensive farming are properly managed. There is some controversy in respect of environmental impacts, though, since many feel that with less surplus cash available within agriculture, and with conservation

Figure 1.3 'No it's not the grain store . . . It's our harvest festival.'
Source: *Farmers' Weekly*

being essentially a voluntary act on the part of the farmer at present, fewer conservation works would be undertaken.

At the same time as price restraint is being considered as a long-term solution to surpluses, a number of other short term measures are being introduced. As a second means of restraint, quotas, particularly for milk, are intended to curtail over-production in the dairy industry, but they have not achieved this to a sufficient degree and in many parts of the EC they have been difficult to police. Some farmers have been avoiding altogether the payment of fines – known as super-levys – for exceeding their quotas.

Partly because of these problems, quotas are not being considered widely as a means of curtailing over-production in other commodities. They are also considered of limited value, by the Countryside Policy Review Panel at least, because they are politically contentious and because they are complex and costly to administer. It is felt that they would be particularly ineffective for cereals, since such produce could be fed to livestock on the farm before it was ever accounted for. Because of this, a third means of restraint, a tax on production, specifically for wheat, was introduced in 1986 – known in the jargon as a co-responsibility levy. Ironically this tax, initially set at 3 per cent of the gross value of output, has been used, almost exclusively, to dispose of surplus wheat.

It is considered by many that, as a fourth means of restraint, directives to divert land from cereals production or to take it out of agricultural use altogether (known as 'set-aside') will have a more effective impact than these production taxes in the short term. Taking land out of agriculture in this way will allow the development of other uses such as recreation, forestry and other economic enterprises (which are discussed later in this chapter) and will be more clearly in the public interest. It has been estimated by researchers at Wye College in the University of London that by the year 2000, assuming current increases in output, about 23 per cent of our agricultural land could be taken out of production to bring supply and demand into balance. This figure could be higher if the lowest grades of land only were 'set-aside'.

Specific measures have, in fact been proposed with the purpose of helping 'set-aside'. Michael Jopling as Minister of Agriculture in 1986 advocated a voluntary system of diverting land, specifically out of cereals production, either permanently or for a five-year period, with compensation for farmers. The 1987 Socio-structures Directive from the EC also contains specific measures for land retirement or diversion. Pre-pension schemes allow payment of farmers who retire early as long as their land is taken out of the production of food for sale. The 'Reconversion and Extensification' scheme gives aid for farmers to move out of surplus products. The whole of the process, however, is not without its critics since many would question the logic of taking land out of agriculture to solve a food surplus problem that has essentially been caused by too much in the way of capital inputs, particularly nitrogenous fertilizers, a point we pick up in chapter 2.

Alongside these 'set-aside' measures, many individuals and institutions are actively researching into alternative agricultural crop production, a development that we return to later in this chapter and in chapter 5. Generally, though, the notion of taking agricultural land out of food surplus production has acted as the principal spur to the development of many other rural activities that form the basis of other sections of this book. The 1987 Circular *Development Involving Agricultural Land* (16/87) was most controversial in the development of proposals in this respect.

The government and others, then, have begun to promulgate measures that will reduce outputs of surplus agricultural products. In this respect, price controls, quotas, taxes on production and agricultural 'set-aside' seem to be the principal preoccupations. More distant, but still under active consideration, is the notion of actually reducing the inputs (other than land) to agriculture in a number of ways. Both Britain and the Commission of the European Communities have proposed a reduction in fertiliser inputs to agriculture to reduce output and diminish pollution. The Countryside Policy Review Panel suggests a European-wide fertiliser levy to achieve this, the revenues from which could be used for further pollution control measures. Similar proposals are suggested for pesticides.

More generally, low input farming is being encouraged with the growth of interest in organic foods and 'conservation grade' produce – the number of organic producers doubled between 1984 and 1986. Environmentally Sensitive Areas, however, are new statutory designations which themselves

Figure 1.4 Organic strawberry growing in Cornwall. This type of agricultural activity is increasingly being encouraged by government policy.
Source: Farmers' Weekly

encourage lower input farming, and we return to these later in this chapter. As an extreme measure, the Countryside Policy Review Panel proposes leaving certain tracts of land with no inputs at all – allowing the land simply to find its natural vegetation climax.

Whether reducing agricultural surpluses will be brought about in the shorter term by constraining outputs or by lowering inputs to agriculture, longer-term solutions are now considered by the government in their *Farming U.K.* document to hinge on fundamental changes in the *attitudes* of farmers. Farmers must now look for alternative means of making a living, either through food products that are not in surplus, or through non-farming means altogether. In the encouragement of this notion, the government has introduced a number of measures to relax restrictions on development in the countryside, and is beginning to offer incentives in the development of alternative rural enterprises. These policy developments are discussed in this chapter.

It is also felt that farmers must now accede much more to consumer concerns about diet, pesticides, other food residues, such as steroids, animal welfare and so on, if they are to cater for an effective demand that will allow them to earn a living. They must also become more aware, along with the food and drink industry more widely, of the potential of the marketing of food products. To this end, a further government policy has been to set up the 'Food from Britain' organization in 1983.

Policies and proposals: developing forestry and woodland

Because, unlike food, we import the majority of our timber into Britain, the government, in its document *Farming U.K.* proposes to continue its policy of encouraging the expansion of traditional forestry. Much of this can be done with relative ease since around half of the productive woodland in Britain is actually owned by the government and administered by its agent, the Forestry Commission. The history of the public production of timber has long been one of expansion which changed, in principle at least, only with the passing of the 1981 Forestry Act. This did not seek to arrest forestry expansion, but rather aimed to shift the responsibility for new planting onto the private sector. It encouraged the Forestry Commission actually to *sell* land – a policy that so far has met with little success because of the limited economic potential of private forestry, which we consider below.

The onus of expansion in practice at present, therefore, rests with both the public and private sectors. The means of public support for private forestry were reformulated in 1981 under the Forestry Grant Scheme which itself was supplemented in 1985 by the Broadleaved Woodland Grant Scheme. Both of these define specific grants and tax arrangements for private forestry as an encouragement to plant trees, and to compensate for the long investment period required to allow trees to come to maturity. In the Broadleaved Woodland Grant Scheme particularly, environmental and access conditions are required before grants may be issued. The Countryside Commission, too, has available an amenity tree planting grant for shelter belts, small woods and so forth, of which both private woodland owners and indeed local authorities may take advantage.

In developing these initiatives further, the government has introduced in 1988 a 'Farm Woodlands Scheme', that will be designed specifically to help reduce agricultural surpluses by encouraging forestry through the existing farm support mechanisms. These will include annual payments to farmers in the early years of tree planting to compensate for the long gestation period of the crop. The Countryside Policy Review Panel, as well as the National

Farmers' Union and Rural Voice, a collection of voluntary sector organizations, welcomes this notion of a 'management allowance', but stresses that it should not unduly favour the richer farmer in the way that tax concessions under the Forestry Grant Scheme have favoured the larger forester.

This Farm Woodlands Scheme is closely related to the need to stimulate existing woodland management. This can be enhanced by, for example, developing local markets for wood products and by innovative approaches to silvicultural schemes. Government funded experiments are currently underway to attempt to develop self-funding woodland management cycles on farmland. These are taking place both in the West of England (Project Sylvanus) and in Wales (Coed Cymru). The Countryside Policy Review Panel also considers that innovations in the ownership of woodlands – community, parish, co-operative ownership and so forth – would allow new forms of management to take place, in the development of 'community woodlands'.

Despite these proposals, there are commentators who consider that such forestry expansion, either to save on timber imports or to assist in the curtailment of food surpluses is economically very questionable. Pressure groups such as the Ramblers' Association in their 1982 booklet, *Afforestation; the Case Against* have argued against the expansion of forestry on economic grounds. The Forestry Commission has always had a notoriously low rate of return on capital (currently around 3 per cent) and actually shows annual losses on its revenue accounts.

The fragility of these narrow economic parameters has been recognized by successive governments. In 1972, the Treasury produced a cost-benefit study on forestry that clearly indicated that broader social benefits associated with forestry would have to be stressed if expansion was still to be an objective of public policy. The ensuing 1972 White Paper *Forestry Policy* emphasized the importance of the multiple use of forests.

More recently, the economics of the forestry sector has come under scrutiny again. In 1986, the National Audit Office, an independent body set up to review the economic efficiency of a number of public organizations, commissioned a study, *Forestry in Great Britain* which criticizes the Forestry Commission's accounting procedures and again emphasizes that the Commission's rate of return on capital is still unacceptably low by public accounting standards. It also questions the real worth of some of the 'public' benefits of forestry such as employment creation and recreation potential. It considers that there are no economic benefits from new planting arising from balance of payment considerations or strategic arguments, and concludes that the Commission lacks any clear objectives to help overcome these problems.

The Auditor General presented these findings to the House of Commons in December 1986 in the report *Review of Forestry Commission Objectives and Achievements*. Here, he criticized the multiple-objective nature of forestry, claiming that there were simply too many objectives for the Commission to achieve properly, and stated that the economic case for forestry could be made only where planting could take place on better quality land, currently used in agriculture.

The House of Commons Public Accounts Committee, the all-party watchdog on public expenditure, responded to this report in April 1987. It expressed concern that many of the Forestry Commission's objectives could not be made more tangibly accountable and called for a greater degree of quantification in the Commission's activities. The low rate of return of forestry could possibly be justified on job creation grounds, but really more information was required about this. More generally, the Committee was left needing to be more constructively reassured about the economic value of forestry.

It seems, then, that whilst the government currently is seeking to expand forestry in Britain, bodies such as the National Audit Office and the House of Commons Public Accounts Committee express deep concern about the economic means of so doing. The possible economic marginality of forestry has led other bodies to look again at forestry potentials outside of timber production.

The Countryside Policy Review Panel suggests that the public benefits of timber, and its integration with farming should be emphasized. In pursuit of these public benefits, the multiple-purpose nature of forestry – particularly its recreation, employment and environmental potentials – should be stressed. In environmental terms particularly, it considers that strategic regional amenity planting plans should be developed, with particular stress on on-farm and urban fringe planting. These could be combined with industrial forestry planting strategies and both might appropriately be produced by County Councils.

Such strategies, which would be backed by a system of planting licences issued by the Forestry Commission, are seen by many as a compromise to requiring full planning control over forestry. This proposal is a closely debated one – the Countryside Commission for England and Wales, for example, favours it, whilst the Countryside Commission for Scotland does not – but it is one which will perhaps most readily lend itself to a compromise solution. Planting licences would be exactly analogous to felling licences which currently exist. It would seem then that the arguments concerning forestry as a significant alternative land use to agriculture in the countryside are finely balanced ones. We will return to look at these in more detail in chapter 3.

Policies and proposals: land use, planning and development

The decline in the pressure to take agricultural land for development that we mentioned in the introduction to this chapter as being one of the prime causes of a substantial rural policy rethink, has coincided with greater levels of productivity on agricultural land through increased capital inputs, enlarged field sizes, improved plant and animal strains and so forth – all factors that have led to our food surplus problem.

These productivity increases have made it less essential to keep land in agriculture anyway. The reduction in pressure on agricultural land itself has been influenced, with varying degrees of success, by a number of government initiatives. The Derelict Land Grant has allowed idle land, particularly in urban areas, to be reused for development. Some of this land has even been restored to agricultural use, causing a net gain in productive farmland.

A second government initiative which was fully operational by 1983 complemented the Derelict Land Grant in allowing under-utilized land to be more fully exploited. This was the introduction of the Land Register of publicly-owned land which identified significant tracts of public lands that were suitable for development. More generally, as a third initiative, Urban Development Grants for small areas and Urban Regeneration Grants for larger ones, have been introduced with the intention of helping to renew decaying urban areas, and the Urban Development Corporations have done much to channel development pressures into the hearts of our cities.

These grant aid policies essentially concerned with steering the location of land uses particularly to underused areas, have slackened the pressures for development in certain parts of the countryside. In other high quality areas,

Figure 1.5 Amenity woodland. Current proposals are seeking to integrate this type of woodland with industrial plantations (see figure 3.4) as a means of increasing amenity benefits and reducing environmental impacts. *Source*: Countryside Commission

however, such as Green Belts, National Parks and Areas of Outstanding Natural Beauty, the pressures continue relatively unabated. The further safeguarding of these 'protected' areas forms part of the Circular *Development Involving Agricultural Land* (16/87) and the DoE's Circular of 1984, *Green Belts* (14/84).

In terms of these government land-use policies, then, the DoE's Circular *Reclamation and the Re-use of Derelict Land* (28/85) maintains a commitment to re-using vacant, under-used and derelict land to allow the reduction of pressure for development in the countryside. Outside of the protected areas, the pressure for development is most acute along the M4 corridor and in the general area bounded by Bristol, Cambridge, Brighton and Southampton/Bournemouth, the heart of the *golden horn* that we discussed earlier in this chapter. Within this belt, it is the urban fringe that faces the greatest problems. It is in the fringe that farming is most susceptible to diversification, and that a jumble of land uses is hardest to resist. The Countryside Policy Review Panel feels that it is critical that new policies to protect these fringe areas, particularly in landscape and ecological terms, are promulgated.

To date, a number of initiatives have been undertaken in the urban fringe. We have discussed in *The Changing Countryside* for example, the Countryside Commission's urban fringe experiments and the initiation of the Groundwork Trusts. These Trusts are to continue, with eleven in operation in late 1987 and another six proposed, but both the Trusts and the experiments are dependent on voluntary good will, and are restricted to being responsive to problems as they arise.

What is now required, suggests the Countryside Policy Review Panel, are more formal urban fringe policies that can become part of the statutory local planning process, co-ordinated in one national strategic urban fringe policy perspective. These policies should be management-based placing priority on agricultural land retention, the development of amenity woodland and new access opportunities – for example the development of 'new commons'.

Generally, then, land-use pressures in the countryside as a whole have declined, although the urban fringe remains a critical development area, and positive policies for its landscape and environmental maintenance seem likely to be a public priority. Such policies are likely to take place alongside more general changes to the town and country planning system, which itself has undergone a significant overhaul since 1985.

Many new characteristics of the system have, or indeed are already having, a significant impact not only on the urban fringe, but also on the deeper countryside. This phase of policy reformulation was initially spurred by the government's 1985 Command Paper, *Lifting the Burden*, which sought to simplify the planning system. It has continued, most significantly, with the more recent 1987 Circular, *Development Involving Agricultural Land* (16/87).

It is this 1987 Circular that, for the first time, gives formal government sanction to the need to move away from agricultural land protection as the general tenet of the planning system. The Circular maintains that the rural economy needs to diversify and to create new job opportunities as long as due attention is given to the protection of the rural environment. Economy, environment and agricultural land loss, rather than the third factor alone, are now to be considered together in making development decisions about rural land.

Reflecting the sentiment of *Lifting the Burden*, planning authorities under the Circular 16/87 now have more freedom in making decisions about 'greenfield' development. They are no longer required to consult MAFF over all developments involving agricultural land, but just developments concerning grades 1 and 2 land. It is this land that the Circular wishes to see retained in agriculture, on the basis that it is the most valuable agriculturally, and that its use in other forms of development would be essentially irreversible.

As well as an easing of development pressures on agricultural land, a shift in the emphasis of agricultural land protection, and the greater autonomy of local authorities to make planning decisions concerning agricultural land, the planning system is also to have an instrumental role to play in the economic diversification of rural areas. This will be considered in the next section of this chapter and in chapter 5. Economic initiatives in rural areas are to be encouraged along the lines of the 1984 DoE Circular, *Industrial Development* (16/84) which maintained that:

> In rural areas provision should be made, appropriate to the needs of each area, for individual development which can be accommodated without serious planning problems. Many small scale industrial activities can be fitted into rural areas, providing much-needed local employment opportunities and helping them to retain a working population.

Housing too, is to be subject to more liberal policies which will greatly assist new housing allocations, particularly in the shire counties of lowland England where pressure for new housing is greatest. But we cannot assume this will automatically lead to the development of housing for the poorest members of the community. The 1987 Circular *Development Involving Agricultural Land* (16/87) seeks to improve opportunities for housing development along the lines of another 1984 DoE Circular, *Land for Housing* (15/84). This latter Circular encourages the development of housing in smaller towns and villages as a means of helping to sustain smaller rural communities. Emphasis is placed on the sensitivity of design and layout.

Figure 1.6 Rural starter homes. Relaxing planning policies will not necessarily lead to more housing for lower income groups.
Source: ACRE

The DoE's statement, *Rural Enterprise and Development* backs up this more positive approach to rural housing, stressing that landscaping and low densities will be important for rural areas and also suggesting that the development of low cost housing and housing developed by rural housing associations will be important elements of future policy, although it stops short of saying how this might be achieved.

These types of initiative have been enhanced by the Rural Development Commission through its sponsorship of the National Agricultural Centre Rural Trust which aims to develop housing associations to cater for low-cost, local needs housing. In addition, the Rural Development Commission has made funds available to counter the inherently higher costs of rural housing schemes. The Countryside Policy Review Panel suggests that the government should introduce changes in the housing cost yardstick to enable the Housing Corporation to develop a specific rural programme.

The re-use of existing buildings in the countryside also has been given official encouragement in a DoE 1986 Circular (2/86) where it is stated that the presumption should be in favour of the redevelopment of existing buildings, even in Green Belts and other protected areas, where environmental sensitivity can be accommodated by conditions attached to planning permissions.

As well as these particular types of development in the countryside, Government proposals also have extended to fundamental changes in the planning process itself. In September 1986, a consultation paper was produced entitled *The Future of Development Plans*, which proposes replacing the structure and local plans that constitute the development plan for an area, with a single tier system of plans to be produced by the districts now responsible for local plans. Counties, formerly responsible for the preparation of structure plans, would be left to revive a form of regional planning and to provide policy statements on individual topic areas. Structure plans would be phased out altogether.

Generally, it appears that the town and country planning system is gearing itself up to accommodate a new era of rural change. Protection of certain sensitive areas is to continue, but a number of measures have been introduced to liberalize some forms of economic development seen by many as essential if agriculture is to diversify away from food production successfully.

Policies and proposals: diversifying the rural economy

We mentioned in the preface to this volume, that the 1986 Agricultural Act laid new duties on Agriculture Ministers. These changes are quite important, so it is worth citing Section 17 of the Act in full.

In discharging any functions connected with agriculture in relation to any land, the Minister shall, so far as is consistent with the proper and efficient discharge of those functions, have regard to and endeavour to achieve a reasonable balance between the following considerations:

(a)   the promotion and maintenance of a stable and efficient agriculture industry;
(b)   the economic and social interests of rural areas;
(c)   the conservation and enhancement of the natural beauty and amenity of the countryside (including its flora and fauna and geological and physiographical features) and of any features of archaeological interest there;

and

(d)   the promotion of the enjoyment of the countryside by the public.

We have already discussed a number of new measures that are being considered by the government and others in terms of maintaining a stable and efficient agriculture. From what has been said above, you will appreciate that it is the stability of the industry in the face of food surpluses that is the major preoccupation of the government in pursuit of objective (a).

In addition to this, however, it is recognized that the maintenance of a stable and efficient agriculture will require special assistance to poorer areas, particularly in the mountains and hills. The Less Favoured Areas Directive was therefore extended in 1984, so that it now embraces some 53 per cent of the land area of the United Kingdom. As well as providing more favourable support for agricultural production (such as Hill and Livestock Compensatory Allowances, or headage payments) these Areas are eligible for special assistance in farm diversification, particularly into the fields of tourism and craft industries. Environmentally Sensitive Areas also require this special type of assistance, to allow conservation objectives to be pursued. We return to these later in the chapter.

It is this idea of farm diversification more generally that is interesting the government in pursuit of objective (b) – developing the social and economic interests of rural areas. Obviously, on-farm diversification is only part of this particular objective, but we will examine this first before going on to look at rural diversification outside of agriculture. We will leave the appraisal of objectives (c) and (d), to the final parts of this chapter.

Farm diversification itself formed part of the Ministerial statement made by Michael Jopling in February 1987 in introducing the draft circular *Development Involving Agricultural Land*. He maintained that under Section 22 of the 1986 Agriculture Act, grant aid would be given for ancillary businesses on, or adjacent to, farms. This was to include activities such as adding value to food products though processing, recreation and tourism, and on-farm marketing. A scheme to administer this – the Farm Diversification Scheme – targeted at small and medium sized farms and specifically for farmers and their workers, was introduced in September 1987.

The scheme has two components: grant aid for capital improvements and assistance for feasibility studies and for marketing. The capital grants are subject to the farmer producing a satisfactory improvement plan and to a maximum of 25 per cent of the cost of the improvement. This is considered sufficient to allow diversification, but not sufficient to encourage un-economic investment. Feasibility studies can attract 50 per cent of their cost, and marketing projects 30 per cent. MAFF's Socio-economic Advisers are closely involved in the running of the scheme. No overlap is allowed between this, and other diversification schemes, which we introduce below.

The Countryside Policy Review Panel goes into a little more detail about the range of farm alternatives available in this climate of diversification. For example, it suggests that there is now a potential for farmers to become involved in the supply of machinery and skills for new economic developments in the countryside and public works contracts. New crops also have some potential, particularly proteins and oil seeds. Moreover conservation itself can become a 'crop', particularly if structural support for conservation to farmers is developed as we shall see later in this chapter. The marketing of foods and regional products can be much more fully developed as an on-farm activity. Finally non-agricultural enterprises on farms could be evolved, particularly in terms of recreation, manufacturing and service enterprises.

Agricultural dependence on multiple incomes seems then increasingly

Figure 1.7 Wild flowers on the farm. This crop has a commercial value for its seeds, as well as enhancing the landscape. *Source*: *Bournemouth News*

likely, but such diversification will need to be encouraged through favourable borrowing rates and fiscal concessions to farmers. Given such support, the proposed Farm Diversification Scheme will undoubtedly help these farm alternatives to be initiated. The objective of developing the economic and social interest of rural areas in the 1986 Act, however, has also been pursued within agriculture with the introduction of a 'Farming and Countryside Initiatives' scheme in 1986, devised by no less than eight government ministries and quangos. The scheme is designed to help farmers and others generate work for the long-term unemployed through the Manpower Services Commission's Community Programme Scheme, although some adjustments to it will be necessary as the Community Programme is superceded by the Adult Training Programme and the Manpower Services Commission becomes known as the Training Commission, on 1 September 1988.

In this scheme it is the intention that rural communities themselves should have the opportunity to develop their own work programmes and that they should principally be of community benefit rather than private gain.

Particular encouragement is given to projects that help to develop local amenities, recreation and tourism, conservation and facilities for small starter businesses. This 'Farming and Countryside Initiatives' scheme thus acts as a bridge between farm and non-farm diversification in the country-side.

The government and others also, then, have an interest in broadening the economic base of the rural economy outside of agriculture. Although this interest has come as outlined in the introduction to this chapter, from the existence of food surpluses, growing rural populations and increasing pressures for development in the countryside, we must not overlook the underlying structural problems of the rural economy that such diversification also will assist. Thus, despite prevailing rural repopulation there are still many areas of remote rural Britain suffering significant population losses. In many more areas, too, there is a continuing erosion of rural services. And although often hidden by population densities, rural unemployment can be particularly severe in the face of a continuing decline in primary sector industries. We return to this theme in chapter 8.

Thus, a large number of factors combine to give rural diversification real potential. In addition, much new industry is seen as being environmentally 'clean' and rural locations for such developments are coincident with the places that appear residentially popular. The Development Commission and its agency, the Council for Small Industries in Rural Areas (CoSIRA) have traditionally had a central role to play in the development of rural enterprise in England. Both organizations, were merged in April 1988 and their regional organization was strengthened.

A significant rural enterprise policy introduced in 1984 was that of the Rural Development Area. This is an area in which special assistance is available for development through the Rural Development Commission, consistent with the implementation of a Rural Development Programme. This programme is a non-statutory document drawn up by the Rural Development Commission, local authorities and other agencies in pursuit of both economic and social development.

The Countryside Policy Review Panel would like to see this type of strategy developed further, to embrace all rural areas and indeed in February 1988, supplementary payments were made to these areas by the Development Commission. The Review Panel is proposing the introduction of Rural Development Strategies, produced by county councils, to embrace economic and social objectives as well as environmental and recreation objectives, and policies for the use of rural land. The DoE could produce guidelines for such strategies, and planning authorities adopt more of an enabling than a regulatory role. A variation of this kind of approach is, in fact, being considered by the DoE as we shall see in chapter 4.

Enterpise assistance available generally for rural areas in England has been re-expressed in the Development Commission's 1987 statement, *Action for Rural Enterprise*. Catalytic services offered by the Commission include the provision of, and assistance with, new premises, many being made developable as a result of the 'liberalizing' of planning controls. They also embrace advice to rural businesses, finance in the form of both grants and loans (where tourism is given a particular priority in Rural Development Areas), training (for which eight different national schemes currently exist) and marketing.

Action for Rural Enterprise also has specific guidance on the development of a number of separate economic sectors in the countryside. These include manufacturing (particularly high technology), food processing, timber products, service industries, tourism and leisure, retailing (including

Figure 1.8 Commercial deer farming at Woburn. A popular and successful alternative livestock enterprise for farmers seeking to diversify.
Source: Farmers' Weekly

pick-your-own), transport, forestry, fish farming, alternative crops and livestock, and organic farming.

The apparent success of enterprise initiatives in rural areas has led the DoE, in their document *Rural Enterprise and Development*, to extending four particular schemes. These have entailed, firstly, the introduction of new advisers for CoSIRA to work closely with Local Enterprise Agencies (essentially private sector partnerships that enjoy certain fiscal concessions from government) in developing programmes of rural diversification.

Secondly, extra assistance has been ear-marked for tourism development, particularly for the conversion of farm buildings into accommodation units and the English Tourist Board and the MAFF have been given specific advisory responsibilities in this respect. Thirdly, extra grant-aid has been assigned to the marketing of rural products, and finally new private sector partnerships, closely related to Local Enterprise Agencies, are being developed for the purposes of job creation.

In addition to these initiatives, the DoE also is keen to further involve the voluntary sector in the development of rural enterprise. To this end, the Rural Community Councils have enjoyed continuing support from the Rural Development Commission as bodies co-ordinating voluntary and community activity at a county level. At a local level, parish councils have been important in the development of both social and economic initiatives. The DoE recently has welcomed, with further support from the Rural Development Commission, a body specifically to co-ordinate the work of the Rural Community Councils at a national level – Action For Communities in Rural England (ACRE). Finally, MAFF's advisory service, is soon to be offering grant-aid for specific farm diversification schemes.

Policies and proposals: promoting recreation and tourism

In all of the policies and proposals that we have discussed so far, from measures to curb agricultural over-production and develop forestry, to reforming the development process and diversifying the rural economy, recreation and tourism continually surface as 'desirable' things to develop from both a public and economic point of view. From these origins, proposals for the development of rural leisure are often rather nebulous, despite the fact that promoting the enjoyment of the countryside is one of the new duties of agriculture ministers under the 1986 Agriculture Act. Although often mentioned as new elements to policy, rarely are recreation and tourism considered a paramount priority. So what types of proposals are positively being considered for these sectors?

The growth in rural leisure participation of the 1960s and 1970s has now stabilized somewhat and although the vast majority of people still enjoy 'passive' recreation in the countryside, there is a discernible trend towards more active 'sporty' types of pursuit. In this context of the importance of rural leisure, the Countryside Policy Review Panel feels that more resources should be made available by both central and local government to protect the opportunity for rural enjoyment and that this priority should match that of both food and timber production.

We can see from the preface to this volume, however, that during three successive Conservative administrations rural leisure time has never attained the political importance of countryside conservation. There have been few government reports concerning rural leisure during the 1980s and approval of such statements of intent as the Countryside Commission's *Access Charter* have involved the government in no resource commitments. The only significant legislation has been for Scotland in the 1981 Countryside (Scotland) Act.

Some commentators have considered that this low priority for rural leisure may be due in part to the fragmented responsibility for it in Government. Juristriction over rural tourism shifted from the Department of Trade and Industry to the Department of Employment in 1984 and more generally, there are countryside recreation responsibilities in MAFF and the DoE, but also interests on the parts of Ministers for the Arts, Health, and Education and Science. The Countryside Commission and the Sports Council too, have been known to dispute who should be in charge of the most popular sport of all, walking in the countryside. We discuss this organizational complexity further in chapter 6.

Within the Countryside Commission in particular, the opportunity to enjoy rural leisure is a clear policy priority. Its 1987 consultation paper, *Enjoying the Countryside* for example was part of a wider policy review process being undertaken by the Commission, known as *Recreation 2000*. The consultation paper provided a direct input into a Countryside Commission recreation policy launch in September 1987 which comprised two main statements – *Enjoying the Countryside: Priorities for Action* and *Policies for Enjoying the Countryside*. In all of these documents, the Commission sees itself as having a much more promotional role than hitherto was felt appropriate under its 1968 Countryside Act terms of reference.

It now proposes to expand its information and interpretation facilities and make them more readily available to the public. This is a proposal endorsed by the Countryside Policy Review Panel, which sees these services being offered in schools as well as on site, as part of a wider process of environmental education. It is proposed that more use is made of the media, particularly local radio, and that the Commission's *Access Charter*, approved by the DoE in 1983, is marketed more widely. Increasing countryside recreation awareness will also entail the establishment of countryside bureaux in larger towns and information packs to be distributed through pubs and shops.

Enjoying the Countryside also has proposals for specific groups; for example, the handicapped, those without cars and even those who have never really been to the countryside before. A positive approach to marketing is again seen as having a role to play here, and specific recreational transport facilities are proposed.

The notion of targeting specific groups in the encouragement of recreation and tourism reflects a recurring theme in the development of rural leisure policies. A report was produced in 1983, entitled *Leisure Policy for the Future* which was signed by the Chairmen of all of the government quangos that have a responsibility for leisure activities. This called for a re-orientation of public leisure policies towards helping the deprived and disadvantaged. This notion was supported and extended by the Association of Metropolitan Authorities, in their 1986 document, *Leisure Policy Now*.

In terms of planning for recreational land uses, the Countryside Commission through both its Countryside Policy Review Panel and its report *Enjoying the Countryside*, has specific recommendations. In line with its proposals for forestry, the Panel sees the need for strategic regional plans produced by local authorities that identify a coherent system of recreation sites and facilities, catering for a variety of needs. This could be done with the help of the Countryside Commission's research facilities. The Commission too, in its policy document *Enjoying the Countryside* is asking highway authorities with significant countryside interests to adopt a new approach and prepare rights of way strategies.

The development of such a system of recreation sites and facilities could, in the future, be made more flexible in England and Wales, if recent Scottish designations are more widely adopted. The Countryside (Scotland) Act of 1981 introduced a new designation – the Regional Park – which, if adopted in England and Wales, would provide a useful supplement to the existing designations of country parks, picnic areas and transit caravan sites.

Figure 1.9 Access to the countryside. The Countryside Commission is seeking to improve the definition, maintenance and management of public rights of way.
Source: Countryside Commission

Figure 1.10 Bunkhouse barns. One of the Countryside Commission's solutions to cheap overnight accommodation and a means of generating farm incomes from recreation.
Source: Countryside Commission

The Countryside Policy Review Panel also feels that the identification of 'open countryside' where people can enjoy a 'wilderness experience' should be made more clear. At a more local level, it feels that public footpaths should be promoted more positively and that access to rural land should be made less intimidating.

In *Enjoying the Countryside*, and the two subsequent policy statements, the Countryside Commission sees making the rights of way system properly available to the public at large as its biggest challenge. It aims to have the 120,000 miles of rights of way legally defined, properly maintained, well publicized, and available for use by the end of the century. It also proffers conciliatory proposals such as the development of local liaison groups of both landowners and users, and suggests that local authorities should review how they manage their rights of way responsibilities. Rights over common land too, would be safeguarded with the use of access agreements where necessary.

In fact, improving access to common land has been a preoccupation in three successive Conservative administrations. From 1979 to 1985, for example, Private Members Bills concerning common land were laid before Parliament each year, but none of them received Royal Assent. For example in 1983 the Access to Commons and Open Country Bill and the Walkers (Access to the Countryside) Bill were both introduced in the House of Commons, but received insufficient support from government to reach the statute book.

In terms of particular facilities, the Countryside Commission proposes to introduce a further five Long Distance Routes for walking, riding and cycling. Low cost maintenance schemes for sites and paths will also be explored. Country parks also will be assisted in a selective way, with priority being given to accessibility and the needs of the disadvantaged.

All of these proposals, however, are designed to serve the public. It is recognized that the income potential of rural recreation is fairly limited, and that the provision of specialist facilities will become necessary in the generation of any significant degree of rural wealth. The Countryside Commission's *Enjoying the Countryside* however, maintains that the development of a good recreation infrastructure, will allow other facilities to become profitable. The development of tourism facilities in particular is likely to generate more employment and income than most other forms of leisure facilities, a view supported by the Cabinet Office's 1985 report, *Pleasure, Leisure and Jobs*. It is perhaps, the limited income potential of recreation that has left it a relatively low priority in rural development, for successive Conservative administrations. Because of this limitation in revenue generation, it will become necessary to develop countryside recreation with the public and private sectors in partnership.

Here, some form of payment to private landowners in exchange for extending public access, seems a likely possibility. This would be in exchange for maintenance and management commitments. All of these developments should take place in tandem with improved information and interpretation activities both in the countryside and in the education system more generally, to engender a sense of appreciation and respect for the rural environment.

Policies and proposals: new conservation measures

The government's 1986 document, *Conservation and Development, the British Approach*, summarizes a wide range of conservation initiatives that are being undertaken in public policy in pursuance of the principles laid down in the World Conservation Strategy. The Strategy was itself initiated at a 1972 United Nations Conference on the Human Environment in Stockholm and was produced in 1980. Many of these initiatives of relevance to the countryside will be familiar to you if you have read chapter 4 of *The Changing Countryside*; many have been discussed earlier in this chapter in a number of sections. This 1986 document does summarize them, however, and presents them in a glossy way, consistent with the government's desire to raise its profile with the conservation lobby.

Developments claimed for the government include, in agriculture, the introduction of experiments to maintain traditional grazing schemes on the marshes of the Norfolk Broads, and the successful development of the Farming and Wildlife Advisory Group, although FWAG was actually set up through a voluntary initiative. In forestry, the document summarizes the Forestry Commission's progress in the appointment of conservation consultants. The Timber Growers Association (United Kingdom), the forestry trade association, has also introduced a code of conservation practice although again this was not directly at the instigation of government.

The working of the 1981 Wildlife and Countryside Act, which was assessed at some length in *The Changing Countryside*, also has been monitored by the Parliamentary Select Committee on the Environment. This Committee suggested a number of modifications to the Act which by and large would afford more effective protection to both habitats and individual species. More importantly it suggested that the principle of paying farmers not to plough and to drain land, which is effectively a payment to do nothing, could not be sustained as a principle of conservation. Some of these suggestions have been incorporated in the 1985 Wildlife and Countryside (Amendment) Act, and in chapter 7 we examine both Acts and the transition between them.

Finally in this review, the government looks forward to likely future developments. Here, it articulates the intention to draw in private sector interests to take a more central responsibility for environmental conservation. It also notes that in 1988 the EC intends to introduce a Directive which will require the explicit consideration of environmental factors in all larger planning applications such as trunk roads and power stations.

In addition to this review, a number of the other documents that we have already considered in this chapter make general statements of intent about the conservation of the countryside. *Farming U.K.* establishes as a public policy of MAFF the conservation role of farmers as custodians and trustees of the countryside. This role is to be supported by positive action from agricultural advisory services. Management agreements under the 1981 Wildlife and Countryside Act are to be continued as is the government support of the County Farming and Wildlife Advisory Groups and the Farming, Forestry and Wildlife Advisory Groups.

As we noted above, the Ministers of Agriculture too, under Section 17 of the 1986 Agriculture Act, are charged with specific conservation responsibilities for fauna, flora, physiographical, geological, archaeological and landscape features. The Countryside Policy Review Panel feels that such conservation measures should have equal priority with food and timber production. This should be achieved through sustainable policies, fully integrated with other policies for agriculture. Such moves also have the support of local authorities, the National Farmers' Union and the Country Landowners' Association who are now calling for more environmentally sensitive land management techniques on farms.

In addition to general encouragement, the Agricultural Improvement Scheme was introduced in 1985 with specific help for conservation measures. It is the scheme under which farmers may obtain support for structural improvement on their farms, and was introduced after a round of structural policy revisions undertaken by the EC, culminating in a new European 'Regulation' (797/85).

Under this scheme for the first time, farmers, and particularly small and medium sized farmers, may be eligible for grant aid for investments in environmental improvement measures and for pollution prevention measures, for planting hedges and shelter belts, building dry stone walls and for constructing footbridges and styles. The Countryside Policy Review Panel welcomes these policy changes, but feels that some of the eligibility criteria particularly relating to the hours worked in, and income derive from agriculture. could be relaxed to allow more people to apply for such grants.

In addition to encouragement and incentives for conservation, the British government introduced a new conservation designation in the 1986 Agriculture Act – the Environmentally Sensitive Area. This stemmed from a specific UK request to include such designations in the EC Structures Regulation (797/85). As we noted earlier, many of the principles of Environmentally Sensitive Areas have stemmed from experiments initiated in 1985 by the Broads Authority in association with the Countryside Commission and MAFF on the Halvergate marshes of Norfolk. As a result, hectarage payments can be made to farm in ways which help to conserve landscape and habitat.

These payments allow farmers to resist the economic pressures for intensification, and as a result, introduce four important features into agricultural support. Firstly, a flexible means of protecting landscape, wildlife, archaeology and so on can be integrated into agricultural practice. Secondly, income support measures are now used to sustain traditional farm enterprises. Thirdly, the Environmentally Sensitive Area places a limitation on the conversion of land to high-yielding food surplus-generating types of

Figure 1.11 The Cambrian Mountains. One of the six original Environmentally Sensitive Area designations. *Source*: Countryside Commission

activity. Finally, the Environmentally Sensitive Area notion generally provides an encouragement to low input, low output farming.

The six Environmentally Sensitive Areas in England and Wales in existence under the 1986 Agriculture Act have been extended to twelve in 1988, but the Countryside Policy Review Panel feels that their objectives as well as their extent should be broadened, for example, to include recreation and access. In remoter rural areas, there are arguments that suggest the objectives of Less Favoured Areas and Environmentally Sensitive Areas should be integrated, to allow a common basis for support.

Issues for the future

Rural policies and proposals are being generated by a large number of organizations, not least the government itself, in many spheres of rural life, from agriculture and forestry through the built environment and rural economy to recreation and conservation. Undoubtedly it is the issues raised in this chapter that have grabbed the rural headlines in the 1980s and it is these issues that provide a basis for evaluating rural policy change in the remainder of this book.

But will these issues, and particularly those relating to agricultural reform still steal the headlines to the turn of the century? Some of you will have noticed, in reading this chapter, stark omissions in the contemporary rural policy agenda. Will the issues of rural deprivation and the loss of rural services come more to the fore in the 1990s? One view of the current rural debate is that it is essentially preoccupied with the lowland parts of Britain. Certainly cereals surpluses, the deregulation of the planning system, the conservation of green belts, recreation pressures and even the 'rape of the

Figure 1.12 Comment on one form of green field development from an unexpected but well-informed quarter!
Source: The Guardian

countryside' do not provide an overwhelming preoccupation for those of us who live in the remoter parts of Britain.

And differing priorities for rural areas are also reflected in different political representation in the near, compared with the remote, countryside. In the 1987 election in particular, the Conservatives did not do very well on peripheral rural constituencies and particularly in Scotland, where unemployment, lack of services and deprivation are important issues. There is a feeling of local alienation and political marginalization in many of these areas, and very little enthusiasm for the unfettered free market.

It is the business of the following six chapters critically to assess the nature of the whole range of rural policies and their implications for our countryside environment. These chapters develop the themes initiated in each of the sections of this chapter, but they are brought together and considered as a whole in the final chapter. There, we will examine their compatibilities, overall shortcomings and means of resourcing and implementation. We will return in particular to the omission of explicit policies to counter rural deprivation and service provision losses.

We hope that through this volume you will be able to develop your own views about the kinds of countryside that we are to realize by the turn of the century and be able to develop your own ideas about their relative potentials to set alongside those that we offer in the final chapter.

Chapter 2

Countering the food surplus problem: too little action for too much food?

Food surpluses and financial shortages

The core of changing rural policies in the late 1980s is clearly agricultural over-production. Though in many respects a very long-standing and deep-seated problem, agricultural surpluses and a resulting level of farm spending well in excess of the resources available, have now grown to the point where they threaten to bankrupt the Common Agricultural Policy (CAP). This budgetary crisis concentrates minds wonderfully and means that agricultural policy reform has come to dominate the political agenda of the European Community (EC), as well as that of the countryside. Agricultural reforms are necessary, not only to balance the budget, but also to give more room to other 'common policies' in the fields of technology, the environment and regional developments which the Community would like to foster. Like the proverbial cuckoo, the CAP has grown into a greedy beast, devouring a large slice of the EC's resources and pushing other policies out of the nest. Moreover, the Treaty of Rome guarantees it a ready supply of resources by classifying farm spending as 'obligatory expenditure'. The European Parliament is expressly forbidden to tamper with the farm budget once it has been agreed.

Reforming the CAP so that farmers receive different signals about what and how much to produce means that some very difficult decisions are having to be made on the nation's farms. These decisions could have a profound effect on the pattern of land use and even the structure of rural society between now and the end of the century. The CAP has been such a powerful motor of change in the countryside in the past that any attempt to tinker with or even dismantle it must inevitably have considerable repercussions. Unfortunately, policymakers still seem more concerned with solving the immediate budgetary crisis than with properly reforming agricultural policy. Most of the reform options which are evaluated in this chapter are therefore necessarily narrow and even crudely conceived solutions to the imbalance in agricultural spending. However, by changing the scale and pattern of farming in Britain, they are also likely to have powerful implications for other areas of rural policy. In an important sense, reforming the CAP and reducing agricultural capacity is the first stage in translating the fruits of past technological progress in the industry into shared benefits for the rest of society. A highly productive farming industry means we can afford to devote

Figure 2.1 Farmers protest at the CAP reforms. Protests like these in Northern France make changes in agricultural support particularly difficult to bring about.
Source: *Farmers' Weekly*

more land to conservation and recreation and even pay some farmers to be managers of countryside rather than producers of food. We discuss the way in which such recreation and conservation initiatives might develop, in chapters 6 and 7.

The common agricultural policy

Renowned for its bureaucratic complexity and phenomenal expense, but also held up as the only truly 'common' policy of the EC, the CAP appears in recent years to have been in a perpetual state of crisis. It may be difficult to believe, but there was a time when the CAP operated smoothly as a self-financing policy. The problem is that the economic circumstances under which the CAP now operates are radically different from those which prevailed when the Treaty of Rome was first drafted in the 1950s. Rapid and widespread technological advances in farming which have produced some dramatic improvements in yields (figure 2.2) mean that the EC can now be fed at reasonable prices by an agricultural industry much smaller than that which presently exists. There have been large productivity gains, with on average one job being lost from European agriculture every minute during the last 20 years, displaced chiefly by technical advances and the substitution of capital equipment for labour, while output has climbed steeply.

In Britain the area under agricultural production has actually fallen as we saw in chapter 1, but increasing yields on each hectare which remains in production have meant that total production has grown very quickly. Modern industrial societies have now reached a stage in their consumption cycles where proportionately less and less of any extra income which consumers receive is spent on food and beverages compared to luxury goods and consumer durables. Given a virtual stability in the demand for food and a continuing increase in production, EC farmers are now producing more of

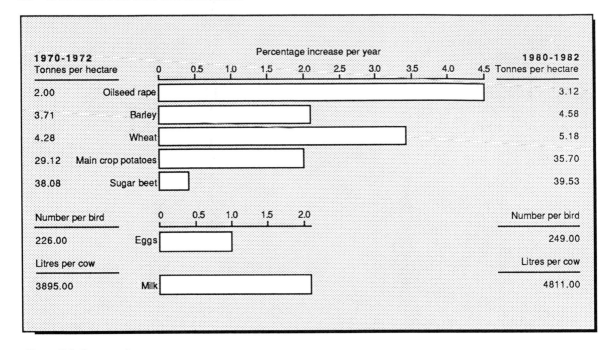

Figure 2.2 Increase in yields in agricultural products. Rising output in a number of commodities during the 1970s has continued unabated into the 1980s and seems set to continue, despite measures to curb over-production.
Source: Ministry of Agriculture, Fisheries and Food

almost all major temperate foodstuffs than can be sold in the market place. Figure 2.3 gives some indication of this by presenting the self-sufficiency ratios for a range of products. In almost every case these are above 100 per cent. It is an unpalatable fact of economic life, at least for farmers, that agriculture is an industry in decline.

Under the CAP, however, where farm prices are guaranteed in order to support farm incomes, farmers are assured a market and a 'reasonable' price for everything they can produce. Now in the early days of the CAP, when levels of self-sufficiency were much lower and the EC had to import some of the food it needed from third countries, it was possible to maintain farm prices and raise revenue by imposing levies on food imports. These prevented a flood of cheap imports from undercutting domestic supplies. So long as production was less than internal consumer demand, the system operated reasonably effectively with levy receipts meeting a large part of expenditure.

This picture changed once self-sufficiency in certain products like milk and cereals was achieved, which was the case in the 1970s. The authorities then had to begin entering the market to buy up excess supplies in order to prevent prices being driven down. These excess supplies had then to be stored or sold on world markets with the aid of export subsidies (which meant that they could be offered at competitive prices). Levy receipts now only cover a small part of the expenditure needed to run the CAP, with member states being called on to allocate an increasing share of their receipts from Value Added Tax (VAT) to the Commission to bolster the EC's 'own resources'. Figure 2.4 shows the size of gross expenditure under the CAP compared with receipts from ordinary and super levies. It also indicates the amount of expenditure that is disposed to guaranteeing prices compared to that spent on guiding structural change.

These last two sources of revenue, ordinary and super levies, are dwarfed by expenditure on intervention. A financial crisis has arisen because spending on agriculture which is undertaken across Europe by a body called

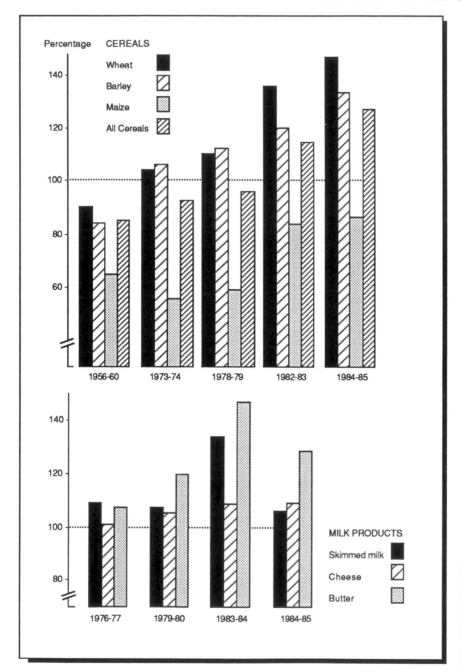

Figure 2.3 Agricultural self-sufficiency in the European Community. Increasing over-production in a majority of products has been a key factor in the argument about the reform of the CAP.
Source: European Community

the European Agricultural Guidance and Guarantee Fund (known univer-sally by its French acronym, FEOGA) has risen by nearly 13.2 per cent a year since 1980. As for the 'community's own resources' these have grown by only 12.6 per cent. It is estimated that in 1987 total farm spending was £2,750 million over budget. Figure 2.5 shows how expenditure to guarantee prices has taken a very large slice of total agricultural spending in recent years. Not surprisingly, the CAP is by far the biggest single item of expenditure in the EC budget, accounting for £9.7 billion of the total £14.5 billion available in 1983. In national terms, Britain made a contribution in that year of £386 million, greater than the contribution it got out of the CAP budget. Taking into account spending by national agriculture departments, spending on agriculture in the U.K. totalled £801 million in 1983, which works out at £4,000 per farmer. It is little wonder that the consumer, contemplating the inflationary effects of the CAP on food prices, is beginning to argue that farm spending has to be cut.

At a meeting of the European Council held at Fontainbleu in 1984, farm ministers also agreed that this state of affairs could not continue and that in future the money spent guaranteeing prices must not be allowed to outstrip the growth of 'own resources'. This marked the first serious political recognition of the need for action, though as we shall see, even this modest commitment to 'budgetary discipline' has been difficult to realise. It has often been remarked that contributing to the costs of the CAP is rather like sitting at a table in a restaurant and agreeing beforehand to split the bill equally with

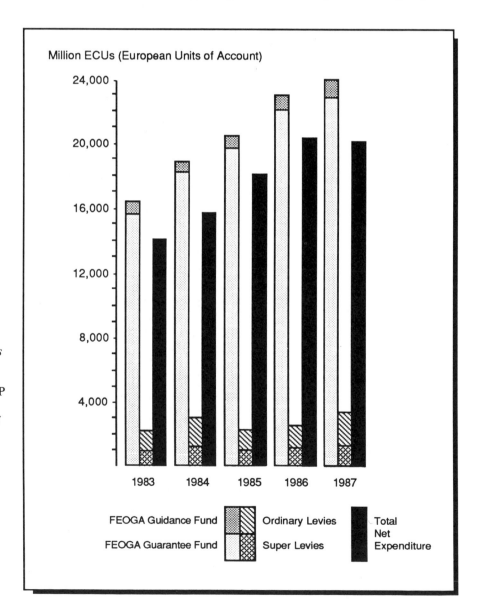

Figure 2.4 Agricultural expenditure and receipts of the European Commission. Expenditure on the CAP dwarfs incomes from it, a situation which largely arises from the amount spent in supporting the prices of agricultural commodities. In this figure the outlay from the FEOGA Guidance Fund plus the FEOGA Guarantee Fund is the total gross expenditure. This minus the levies equals total net expenditure.
Source: Eurostat

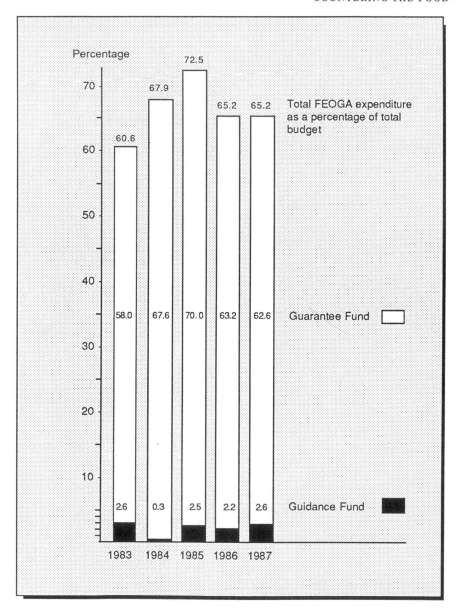

Figure 2.5 Agricultural expenditure as a percentage of the European Community budget. European Community spending on agriculture has never been less than 60 per cent of its total outlay since 1983. Price guarantees make up a substantial proportion of this.
Source: Eurostat

other members of the party. There is little incentive to stint on the dishes ordered because we know that the cost is to be evenly divided. In their positions around this table, farm ministers have been only too willing to keep ordering from the menu. Throughout the 1970s they were able to indulge without worrying too much about the bill; new guests joined the party, including the UK which, as a major food importer could provide more resources. Another factor which postponed the day of reckoning was the rapid expansion of export markets and, with them, the opening up of opportunities for exporting Europe's food surpluses abroad. These operated like a safety valve and meant that, for a while at least, what was a declining industry could behave like an expanding one.

By 1983 a financial crisis finally seemed inevitable. An agreement to raise the proportion of receipts from VAT which member states contribute to EC

Figure 2.6 Reaping the harvest? Whilst the European Council were arguing the case for cuts in cereal production, farmers were still investing in new machinery.
Source: Farmers' Weekly

funds from 1.0 to 1.4 per cent from January 1986 eased the immediate problem of insolvency, though it was recognized that the extra resources which this provided would soon be swallowed up unless farm production could be cut. At the same time, the dumping of European surpluses on world markets had led to friction with other trading nations like New Zealand, Australia, Canada and the United States. The United States threatened a trade war. In 1985 the European Commission issued an important consultation paper entitled *Perspectives for the CAP*, in which it recognized that the CAP was at a crossroads.

Without reform, the paper argued, it would break down, an eventuality that could threaten the cohesion of the EC itself. The problem was expressed in the starkest terms: agriculture, like the rest of the economy, must be exposed to the rigours of the market place and the laws of supply and demand. Resolving the budgetary crisis in any long-term sense required a reduction in non-agricultural capacity. What was less clear in the paper and in the debate which has continued ever since, is precisely how to go about reducing capacity. Which factors are in excess supply: land, labour or capital? It has been argued at different times that there is either too much land or too much capital in the farming industry. The farm lobby has been at pains to present the over capacity issue as a land surplus, arguing that at some future date there will be millions of hectares of surplus land 'searching for an alternative non-agricultural use'. It is no surprise that the lobby is less easily persuaded that their industry is over capitalized or that there are too many farmers on the land.

We will return to the 'too much capital' and 'too much land' issues later in this chapter, but those who subscribe to the 'too many farmers' view argue that the budgetary crisis has arisen in the first place because price support has been used in an inappropriate way to support the incomes of marginal farmers. The solution is to encourage them to leave the industry so that prices can then be brought down without causing too much distress. This, what is in fact a structural solution to the budgetary problem, was first set out as long ago as 1968 when Sicco Mansholt, the then Commissioner for Agriculture, produced his controversial plan to streamline the industry. Mansholt pointed out that many very small producers were often 'locked up' in their farms, running businesses that were unlikely ever to generate a decent income. He quickly realized that attempting to support these producers by subsidizing prices through price guarantees would be self-defeating because it would over-reward the larger and already efficient farmers and lead to the production of surpluses. He proposed paying marginal farmers to leave the land and retire or retrain in industry. This would leave a slimmer industry of large, viable producers who could remain profitable at much lower price levels.

Mansholt's Plan received a predictably hostile reaction, particularly from the powerful farm lobby, who foresaw a loss of members and hence a weaker power base as people were paid to leave the land. Member state governments were also lukewarm about the idea of a muscular structural policy under the direction of the European Commission. They jealously guarded their freedom of action in this area of agricultural policy, fearing a shift away from

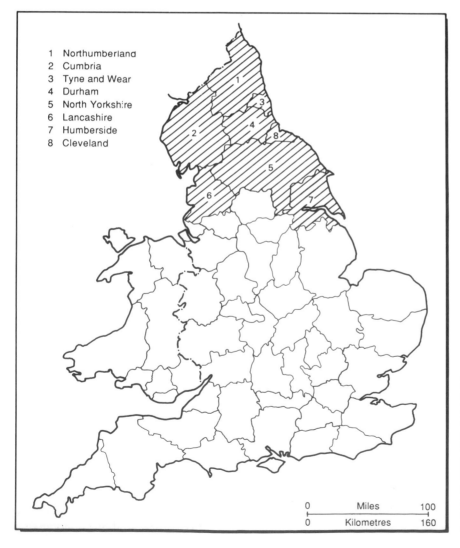

1 Northumberland
2 Cumbria
3 Tyne and Wear
4 Durham
5 North Yorkshire
6 Lancashire
7 Humberside
8 Cleveland

| 0 | Miles | 100 |
| 0 | Kilometres | 160 |

Figure 2.7 Agricultural land surplus – what could it amount to? It has been estimated that about three million hectares of farm land in Britain could be taken out of production to bring food production and consumption into balance. This would be equivalent to the counties identified on the map.
Source: N. Curry

national institutions in favour of supra-national ones. Not for the first or last time, nationalist interests were to block progress towards a common structural policy. Meanwhile, those who stood to gain most from a 'Mansholtian' policy – small farmers and consumers – were too poorly organized to have any influence.

Rather than embark on a long-term programme of structural reform, member states' agriculture ministers emasculated Mansholt's proposals, agreeing in 1972 to a much weaker set of measures. We can get some indication of this by looking at how much less is spent on structural measures (the guidance fund) compared to guaranteeing prices in Figures 2.4 and 2.5. These included a Directive on farm modernization, which grant-aided investment schemes on farms in order to raise incomes to a level which was comparable with those of industry. The emphasis was to be on raising the incomes of existing farmers rather than rearranging the number of farms. An early retirement scheme was in fact agreed, though its impact has been negligible, given inadequate funding and a shrinkage of job opportunities outside farming. Now, with a new round of market policy reforms in prospect, it has been urged that a structural solution should once again be considered. Judging by recent inter-government statements, however, a resuscitation of the Mansholt philosophy is extremely unlikely. These argue against a European agriculture on the model of the United States, with 'large reserves of land and few farmers', stating that a substantial number of farmers need to be kept on the land for social and environmental reasons. Rural communities, particularly in remoter parts of the EC, would be threatened if farmers were paid to leave. This is a deeply entrenched view amongst the member states, especially France and West Germany, the former being concerned about the 'rural desertification' which might result from speeding up structural change. There are some commentators in Britain who take issue with the argument on environmental grounds, pointing out, as we noted in chapter 1, that the evacuation of large parts of the upland countryside would lead to a 'man-made wild' of great conservation and recreational value. Their's is a minority view.

Getting the prices right

Having ruled out a structural solution as a realistic panacea to over-production, policymakers are therefore faced with the need to reduce output on existing farms, using price or quantity instruments to withdraw capital or land from a farming use. Supporters of price instruments believe that market forces should be allowed to determine how much land and capital is withdrawn from the industry. Right-wing political research groups like the Adam Smith Institute have proposed a complete liberalization of farm policy, with all price support being phased out so that prices can fall close to those on world markets. In this scenario, some farmers would undoubtedly be forced out of business as the natural forces of readjustment began to reassert themselves to bring supply and demand into line. Asset values would plummet, forcing vulnerable farmers and their capital stock out of business. The result would be a smaller industry of larger producers and agribusiness-men who would survive by staying on the technological treadmill. Given the level of economic distress and the impact on the rural economy, few regard this as a likely scenario.

A more moderate and a politically more realistic option would be to reduce the level of price guarantees to cut the prices which farmers receive for their production. This policy would solve the surpluses problem by making it less profitable to increase output at the margin, encouraging farmers to switch into different enterprises and extensify their farming methods. Lower price supports would bring immediate savings in expenditure since intervention buying would need to be carried out less often, while export subsidies on produce sold in world markets could also be reduced. Lower prices would obviously benefit consumers and may marginally stimulate demand for food. A low price regime would also have long-term benefits by slowing down the rate of productivity growth as farmers find themselves less able to invest in the latest equipment or machinery.

But how low is low? The extent to which prices, particularly cereal prices, would have to be reduced to cause farmers to reduce output is still a matter for speculation, although it is agreed that cuts would have to be significant if they are to outweigh the effects of continuing technological progress. It has been estimated that a price cut of at least 33 per cent in real terms would be necessary to remove the cereal surplus, for instance. The drawback with more moderate price cuts is that some farmers might then be able to increase production in order to compensate for the squeeze on profits. Most experts agree that this would not happen provided price cuts are repeated each year: efficient farmers would find it progressively less economic to use additional inputs (fertilisers, farm chemicals, concentrates) as the price received on each unit of output falls. However, there may be a time lag of a few years before restrictive pricing can exert its full impact. It must therefore be pursued consistently over a long period of time.

The dilemma for policy makers is that the deep price cuts which would be necessary to provoke a 'supply response' from the largest and most efficient producers, who account for the bulk of the over-production, will at the same time threaten the survival of smaller and more marginal farmers. There can be little doubt that a reduction in farm support would cause genuine hardship for many individual farmers, particularly those who have borrowed heavily in recent years to finance new investment. Farmers' incomes and their ability to service debts are already heavily dependent on state support. For the UK it has been estimated that the total value of price support and all national aids to farming amounted to £1,800 million in 1984, whereas net farming income was just £1,826 million. It has been estimated that something like 10 per cent of farmers could be bankrupted should deep price cuts be carried out.

It is therefore generally accepted that lower levels of price support will have to be accompanied by compensation or income aid schemes directed at the worst affected or most vulnerable farmers. Such an arrangement would mean abandoning price subsidies as a way of supporting farm incomes, restricting them to a role as a safety net only – something which economists have been urging for years. The difficulty for policymakers is in deciding which farmers to support. Having decided which to support the next difficulty is in justifying the decision to those who fail to qualify. There is also the risk that the CAP will be gradually 'renationalized' as price guarantees are reduced and individual member states, anxious about the impact on their own farming communities, introduce their own aid schemes without any reference to EC policy objectives. This would distort inter-European competition and severely threaten the CAP as a 'common policy'.

Recently in the European Parliament, a number of alternative common aid schemes have been canvassed in an effort to prevent this happening and to ensure that income aids are used as long-term investments, rather than merely short-term compensations. One option would be to offer the large number of EC farmers nearing retirement the chance to give up farming in return for an EC 'pre-pension' which would be paid until the normal retirement age was reached. Another would be to target income aids at farmers who would be viable in the long run provided they could be helped

over the period of transition to lower prices. A third option would emphasize the social welfare purposes of income aid payments, directing aid to the most needy farmers regardless of whether or not they can remain viable in the long term.

Recent proposals from the European Commission suggest that income aids will be directed at 'intermediate or middle band farms' which, while facing difficulties in adjusting to falling prices, will be capable in the long run of becoming viable businesses. Payments are also to be made available to pre-pensioners' however, provided they also agree to release the land to younger successors or agree to take some land out of agricultural production. In putting forward these schemes, the Commission has stressed the importance of linking market reform to a stronger socio-structural policy. It argues that as spending on guarantees is reduced, more money should be available to spend on 'guidance' under a structural policy which is designed to assist the process of change and ease the transition to a more streamlined industry. Under a 'mini-Mansholt' policy 'guidance' expenditure under the CAP has been dwarfed by spending on price support under the 'guarantee' section, as we have shown in figures 2.4 and 2.5. The Commission, and many others, including conservationists in Britain, would like to see structural (or 'guidance') policies absorbing as much as 25 per cent of total agricultural spending by the early 1990s. If it can be achieved, this new balance of expenditure would create many new opportunities, not only for helping individual farmers, but also in ensuring that wider environmental and social objectives can be built into the adjustment process. The Environmentally Sensitive Areas programme which is discussed in chapter 7 is a notable example of a policy which will be part-funded from the guidance fund. The idea that farmers' incomes can be supported by paying them to 'produce countryside' is likely to be an important one in future socio-structural policy.

Getting the prices right will not be easy to accomplish, even with schemes to compensate the most vulnerable farmers. The record of farm ministers in agreeing to price cuts is not an encouraging one. After protracted price fixing sessions, ministers typically agree a compromise package which waters down the original Commission proposals and includes special concessionary measures. Many of the 'regionalized' socio-structural schemes, such as the integrated rural development programmes (IRDs) in Ireland, have been agreed as make-weights in the annual price reviews, hastily concocted to secure ministerial agreement with a package of measures. In 1986, West Germany was allowed to extend the coverage of its Less Favoured Areas in return for agreeing the price settlement of that year. Such is the nature of policy-making in a Community of Twelve!

Even where modest reductions have been agreed it has always been possible to offset the effect on national prices by devaluing the 'green' rate of exchange between European Currency Units and national currencies. 'Guarantee thresholds' were introduced for milk and cereals in 1982 in an effort to introduce more rigor into the decision-making process. Under this approach, production thresholds are agreed in advance, with cuts in support prices being triggered automatically whenever annual production exceeds this threshold. In the cereals sector, support prices were to be reduced by one per cent for every million tonnes over the agreed threshold, up to a maximum price cut in any one year of 5 per cent. Unfortunately, farm ministers continued to agree nominal price increases, so the system operated merely to abate these, rather than trigger cuts. Thus, in 1982–83 production exceeded the threshold by a million tonnes, and a price increase for cereals was adjusted from 4 to 3 per cent. In 1983–84 the record harvest of that year produced an excess production over the threshold of

Figure 2.8 Marginal farms in the European Community – a Brecon Beacons example. Reforming price guarantee policies will always threaten to put marginal farmers out of business.
Source: Countryside Commission

*Figure 2.9
Environmentally
Sensitive Areas – the
Test Valley in
Hampshire.* Greater
spending on the
guidance part of
agricultural policy can
better support the
maintenance of
environments such as
these.
Source: Countryside
Commission

*Figure 2.10 A British
foothill in the grain
mountain.* Taxes on
cereal production are
designed to reduce these
Community surpluses.
Source: Farmers' Weekly

8 million tonnes, implying a price cut of 5 per cent. Farm ministers were
unable to agree on this and, after a protracted debate, the Commission itself
was forced to step in and announce a 1.8 per cent reduction in the new
season's prices.

In February 1988 the European Council agreed to implement a system of
budgetary stabilizers, effectively a revamped version of the guarantee
thresholds. Cereal price cuts will now be triggered when annual production
exceeds 160 million tonnes. Although this high production threshold has
been criticized as still being well in excess of consumer demand, the Brussels
agreement, which covers other arable sectors, is certainly an advance on the
previous position.

A greater use of taxes on over-production, known as co-responsibility
levies was also agreed in Brussels as a complement to the price cuts which will
be triggered under the budgetary stabilizer system. It is argued that as well as
reducing production, farmers should be made to bear some of the costs of
disposing of surpluses. In practice co-responsibility levies are little different
in their effects from a cut in the prices which farmers receive, although unlike
price cuts they have the advantage that smaller producers and other
identified groups can easily be exempted from the levy. Where they have
been introduced in the past, co-responsibility levies have been too small to
have any impact on production and have been ostensibly used to generate
revenue to fund the disposal of surpluses or, as in the case of the milk levy
imposed in 1977, to subsidize marketing campaigns to raise milk consumption. They thus tend to be regarded as somewhat cosmetic and short-term
measures which, while useful in easing budgetary constraints, do little to
solve the over-production problem or help consumers and users. There is
also the danger that, under pressure from farming interests, support prices
might actually be raised to offset some of the effects of a substantial levy.
Levies have already been imposed on all but the first 25 tonnes of grain which
is sold to grain merchants by farmers, with the aim of encouraging more use
of grain on farms as animal feed and as a way of generating additional
revenue for the cereals regime. This system came into operation in July 1986
and is already creating administrative problems.

Restructuring output through quotas

Despite the preference of the European Commission and most academic economists for price instruments, the use of quantity controls is still favoured by certain interest groups as a solution to over production. Quotas, nitrogen taxes (reducing capital inputs) and set-asides (reducing land inputs) have all been promoted by the farming lobby who fear the income and structural implications of price cuts. Quotas, for instance, are likely to be far less draconian than price cuts because they merely set a limit on the amount a farmer can produce before being penalized by levies. Alternatively they can be used to put a ceiling on the amount a farmer can produce which qualifies for price support. Price cuts, by contrast, reduce the level of subsidy on *all* units of production. Many farmers can happily produce below quota by making relatively painless adjustments to the way they farm. The introduction of milk quotas in 1984 provoked a quick response from dairy farmers, who culled cows and reduced their use of high yielding feed concentrates to avoid paying the super levy which was imposed by the government on excess production. Research has shown that dairy net farm incomes have actually risen since the quota system was introduced.

The usual objection to extending the quota empire into other sectors is the considerable bureaucracy which this would entail. Imposing a system of quotas on the cereal sector would be an administrative nightmare since, unlike milk, grain is sold to a complicated network of merchants who subsequently sell it to the main users. A large amount of grain which is grown by farmers is fed to livestock. Faced with a quota on their sale of grain, farmers with livestock could be tempted to avoid paying a levy by feeding more of their own grain to their own pigs or cows. There have also been problems in enforcing payment of the levy in certain areas of the EC, as we noted in chapter 1.

Economists dislike the idea of quotas because they restrain efficient producers and lessen the pressures on inefficient ones. If an industry is to remain dynamic and cost-effective, they would argue, it is important that the most efficient firms expand at the expense of the least efficient. Quotas fossilize production patterns because, provided they produce below quota, marginal farmers are assured of continued market support. This objection can be overcome where farmers are free to 'trade' quotas so that, if they wish to expand, efficient producers can add to their quotas by buying from other

Figure 2.11 'Whoever wins gets the milk quota.'
Source: Farmers' Weekly

producers. This last group will then effectively receive a capital sum as compensation for forgoing the right to produce this amount. Governments can also enter this market if they wish to restrict production still further, buying up quotas and then simply refusing to re-sell.

Nitrogen limitation policies

In addition to quotas, there have been proposals for policies which would withdraw certain capital inputs from agriculture directly. Nitrogen is an extremely productive input in modern agriculture. As figure 2.12 shows, an average of 117 kg per hectare/yr was being applied by UK farmers in the early 1980s. Yields are known to be very responsive to additional nitrogen. It should therefore come as no surprise that nitrogen taxes have been proposed as a relatively straightforward way of reducing output by cutting these yields. The tax levied on each kg of nitrogen would act as a disincentive. The drawback with this solution is that the response of crop yields to nitrogen application is so favourable that the tax would need to be very large indeed to have any effect. However, a heavy tax would also reduce farmers' net incomes. Even if it succeeded in reducing usage, it would raise the unit costs of production, subsequently inflating food prices to consumers.

A fertilizer tax could also have a number of distributional effects, penalizing grassland farmers more than crop producers due to the closer relationship between fertilizer use and income for the former group. An alternative way of limiting nitrogen use would be to impose nitrogen quotas on individual farmers (in other words, input restrictions rather than the output restrictions that we discussed earlier). This could be done by issuing farmers with ration books for purchases from licensed suppliers. Non-licensed suppliers would be prohibited. Although the impact of quotas on farm incomes would be much less than that of taxes, the system would be extremely difficult to administer. How are nitrogen quotas to be allocated to farms of different cropping and soil types? How would you stop larger farmers producing their own nitrogen on site, as a means of avoiding the tax? As with the output quota discussed above, there would also be the danger that production patterns would be frozen in time. Again, the free buying and selling of quota might reduce such distortions.

Set-aside and land diversion

Supporters of an American-style set-aside policy for British and European agriculture start from the assumption that it is land which needs to be withdrawn from production. Land budget studies, including those conducted at London University's Wye College, have shown that as many as 3 million hectares of land could be surplus to requirements by the end of the century if productivity continues to improve on the land which remains in farming. Another study suggests that the UK could still be 80 per cent self-sufficient in temperate products while farming only 77 per cent of its present farmed area. These projections have been seized on by some commentators and interest groups to support the idea that the state should be actively paying farmers to withdraw land from agriculture. They believe that such a policy would be highly cost effective since farmers will be willing to set-aside land in return for a government payment which works out much cheaper than the cost of subsidizing the surplus production on the same piece of land.

The notion of withdrawing land from agriculture is actually not as revolutionary as it sounds. As we have already seen, the Mansholt Plan proposed paying farmers to leave the land. It was thought that over 5 million

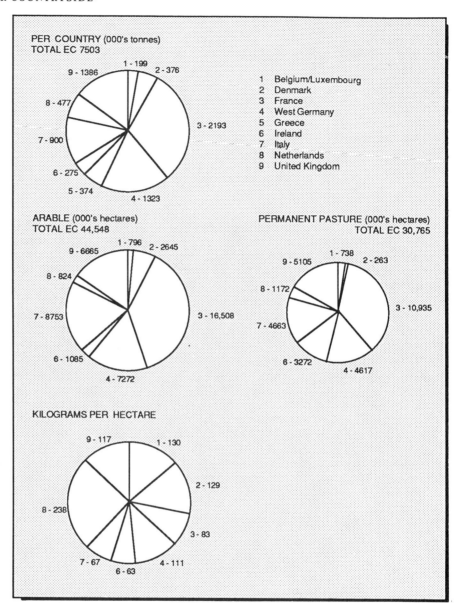

Figure 2.12 European Community consumption of nitrogen, 1981–82. The very high levels of nitrogen use in EC agriculture for crops and pasture are an obvious target when exploring measures to curb agricultural output. *Source:* Eurostat

hectares held by such farmers would have to be redeployed and diverted into forestry or recreation. Mansholt envisaged set-aside as the outcome of structural change, however, not as an end in itself. More recent proposals take their cue from the American experience, where set-aside is pursued as a policy in its own right. Here there have been two rather distinct approaches taken. Under the vast Acreage Reduction Programs (ARPs), which were first set up in the 1930s, large numbers of farmers have been paid to fallow land for short periods to allow supply and demand for surplus crops to be brought back into equilibrium. By contrast, the Conservation Reserve Program (CRP), which has now been set up by the 1985 US Food Security Act, pays farmers to divert land into forestry or grassland for much longer periods. As its name implies, the CRP aims to reduce the soil erosion problem which affects large areas of the United States. Land with an erosion hazard is specially targeted and participating farmers have to have a conservation plan approved before they qualify for land diversion payments. Much of the initial enthusiasm of British conservationists for set-aside is probably due to the expectation that it would be the CRP model which would be implemented here. If this is so, then they have been disappointed.

The influential farm lobby, particularly in West Germany, has been energetic in promoting what might be called full blown set-aside, which has more in common with the ARPs than with the CRP. They envisage a large-scale, but essentially expedient programme of fallowing to reduce surplus production. Supporters point to the American programmes and their undeniable impact in reducing over-production. In return they are reminded of the demonstrable weakness of set-aside as a solution to over-production.

Figure 2.13 American prairie farming. A landscape such as this in south east Iowa will change as farmers divert from cereals production under the US Conservation Reserve Program.
Source: Ron L. Laird, Barnaby's Picture Library

The greatest difficulty lies in administering set-aside schemes and in justifying to the public and farmers themselves the psychologically difficult notion of apparently paying farmers to 'do nothing' (though in practice farmers may be called on to do a great deal in managing the land which is diverted from agriculture). If policymakers are to prevent farmers cheating, perhaps setting aside less land than they are paid to, they will need information about the pattern of cropping on a farm before set-aside is taken up – what the Americans call the farmer's 'base acreage'.

In Europe, the combination of a poor data base and the need to deal with large numbers of farmers, could pose difficulties for set-aside. Even in the United States, where farmers are fewer and information is more abundant, it has been found that the area of surplus crops harvested after a set-aside scheme is in place tends to fall by less than the area which is meant to be set-aside. This is called 'slippage' and has a variety of explanations. Gaps in policing schemes means that there is always some scope for farmers to cheat by diverting less land than they are paid for. At the same time, farmers participating in the schemes have been known to bring new land into production to compensate for the land they have to set aside.

Set-aside is usually voluntary; that is, farmers choose whether or not to enrol land in a scheme in return for payment. This brings further problems. Farmers who resolve *not* to participate may continue to plant more land with surplus crops and further exacerbate the slippage problem. The incentive to 'free ride' on the actions of farmers who *do* participate will be especially strong if farm ministers feel that set-aside justifies them in agreeing support price increases each year. Full-blown set-aside thus has very little to recommend it, despite the support it continues to attract from the farming lobby. Most experts agree that adopting set-aside as an alternative to restrictive pricing, as proposed by the West Germans before the 1988 Brussels summit, would have been a retrograde step. As one commentator has pointed out, once in place, policy makers would always have found it difficult to find good enough reasons to abandon set-aside. Like a drug addict, the farming industry would be hooked on the payments which set-aside could provide.

A set-aside programme which is implemented alongside price cuts is a different proposition. In this role, set-aside payments could be used to speed up the diversion of marginal cereal growing land on particular farms.

Figure 2.14 Hooked on set-aside?
Source: *Big Farm Weekly*

Payments to farmers to set-aside land could also be treated as a form of compensation, though many experts still feel that direct income payments would be preferable. This misleadingly named 'extensification scheme' which was agreed by farm ministers early in 1987, is intended to speed up the process. Under this scheme, cereal producers will receive payments in return for an agreement to reduce their cereal hectarage by at least 20 per cent inside a period of 5 years. Beef producers may also be eligible for payments when they reduce their production of beef by at least 20 per cent. In the case of cereals, farmers will be expected to fallow or afforest the land involved. We will look at the economic wisdom of this kind of move towards afforestation in the next chapter. Set-aside clauses have also now been written into the proposed 'pre-pension' scheme which is intended to encourage farmers in the 55 to 65 age bracket to move out of conventional agriculture. Participants would again have the option of fallowing or afforesting land in order to qualify for annual payments.

Can the CAP be reformed?

The Brussels agreement of early 1988 appears to indicate that the price instrument and co-responsibility levies are to be given leading roles in attempting to solve the over-production problem, though it remains to be seen just how effective the system of budgetary stabilizers proves to be. There is a long history of false dawns in reforming the CAP. The number of Commission proposals which have been shelved or fudged by farm ministers is too great to mention. In view of this, many ask 'Can the CAP be reformed?' Will the budgetary knife turn out to be as sharp as some expect? True reform is limited by the fact that the relevant articles in the Treaty of Rome are unlikely to be rewritten, so the basic goals of agricultural policy as enshrined in Article 39 have to be treated as fixed points.

In addition, commentators point to the fundamentally flawed nature of decision-making in the Council of Ministers, where the need to reach

Countering the food surplus problem.

This diagram explains the effects of all of the measures that we discuss in this chapter, to counter food overproduction, represented in terms of supply and demand functions.

1. WHAT INFLUENCES THE SUPPLY AND DEMAND FOR FOOD PRODUCTS?

The quantity demanded of any food product (Qd) is influenced by a number of factors: $Qd = f(P1, P2-Pn, Y, S)$.

The quantity supplied of any food product (Qs) is influenced by another set of factors: $Qs = f(P1, P2-Pn, f1...fn, G, T)$.

Where:

P1	=	Price of the commodity in question (say wheat);
P2-Pn	=	Prices of other commodities;
Y	=	Incomes of the population;
S	=	Tastes of the population;
f1...fn	=	Costs of the factors of production in agriculture - capital, land, farm labour;
G	=	Tastes or preferences of the farmer
T	=	The state of technology

Figure 2.15 Countering the food surplus problem.
Source: N. Curry

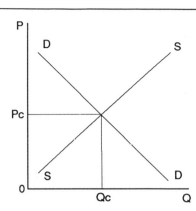

2. MARKET SITUATION OF COMMODITIES IN BALANCE

At price Pc, consumers demand exactly the same quantity of produce as farmers are prepared to supply. This is Qc and the market is in balance.

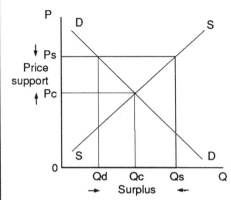

3. MARKET SITUATION OF COMMODITIES IN SURPLUS

If prices are subsidised say by the amount Ps-Pc, the price support for the commodity, farmers are prepared to supply Qs (where Ps hits the supply curve) but consumers are prepared to buy only Qd (where Ps hits the demand curve). This results in a surplus of Qs-Qd. Most of the costs of the CAP are spent in supporting prices in this way (to ensure a reasonable income for farmers) but it does lead to food surpluses.

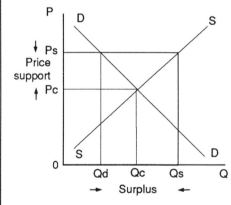

4. OVER PRODUCTION SOLUTION 1:
PRICE RESTRAINT

Mechanically, the simplest solution to the problem of food overproduction would be to reduce the support price (Ps) so that supply and demand were in balance (at Pc). This 'balance' price is likely to be close to the world market price for the commodity. In the short term, this is considered to be a politically diffificult solution because it will mean significant losses to farmers' incomes. When many farmers have borrowed money to buy factors of production (the f1...fn supply function in 1.) these price reductions could make farmers bankrupt. This is because these factors of production influence the quantity supplied.

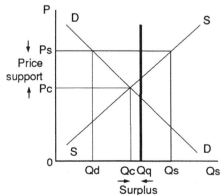

5. OVERPRODUCTION SOLUTION 2:
QUOTAS

Surpluses can be reduced by placing a physical limit on how much farmers are allowed to supply. This is so for milk, where a quota of Qq is imposed. This reduces surpluses (to Qq-Qc) but still allows the farmer to be paid the higher price (Ps) for each litre of milk he produces. His total income will fall, but not by as much as if the price of milk were reduced to Pc.

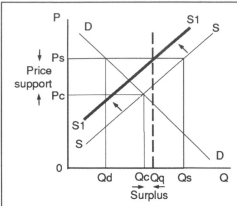

6. OVERPRODUCTION SOLUTION 3:
THE CO-RESPONSIBILITY LEVY

Surpluses can also be limited by co-responsibility levies, or taxes on production. This is so for wheat. A tax on production is essentially an increase in factor costs, which has the effect of pushing the supply curve up to the left (from S to S1). This will reduce surpluses to Qq-Qc with prices to the farmer remaining at Ps. Farm incomes will fall, but not as much as if there was a price reduction to Pc.

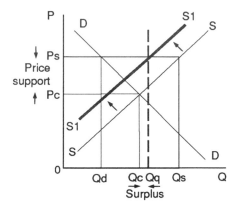

7. OVERPRODUCTION SOLUTION 4:
'SET-ASIDE'

Set-aside entails taking land out of production. This is one of the input factors (one of the 'f's in the supply function) and therefore has the effect of shifting the whole of the supply curve. Because we are taking land out of production, the supply curve moves upwards to the left (from S to S1). This reduces surpluses to Qq-Qc and farmers still get the higher price Ps for the commodity. Their incomes again fall (without any compensatory payments) but not by as much as if prices were reduced to Pc.

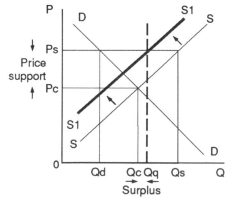

8. OVERPRODUCTION SOLUTION 5:
REDUCING AGRICULTURAL INPUTS

This is essentially the same as 'set-aside' in terms of its economic effects, except that all of the factor inputs (f1...fn) rather than just land are reduced. This is designed to lead to a low input (supply curve moves to the left), low output (reduction from Qs to Qq) agriculture, with much smaller food surpluses (Qq-Qc). The farmer's gross income will be reduced, but because his costs are lower (fewer factors), his net farn income may not be.

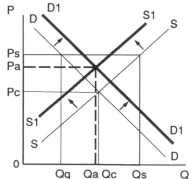

9. OVERPRODUCTION SOLUTION 6:
CHANGING PEOPLE'S ATTITUDE

This is a long-term solution to over production but is concerned to manipulate both the farmer's preferences (G) and those of the consumer (S). This may encourage a move from one product to another, but in our graphical representation of one product only, we have postulated that we have encouraged people to consume more (say through advertising the health properties of the product). You will see in this diagram that this has led to a new equilibrium where supply and demand are both in balance at PaQa.

unanimous agreement on any new proposal means that a single member state can hold up progress indefinitely. Even the modest principle of budgetary discipline has proved difficult to implement, and the guarantee threshold system has not been operated with as much rigour or consistency as it might have been. Until recently, the Commission has been engaged in a perpetual juggling act to maintain the flow of money which will keep the CAP solvent on the one hand, while putting in place longer term measures which will bring markets into a better balance on the other. In the absence of any agreement by ministers to follow these proposals through, the Commission has already embarked on its own set of incremental solutions. It has, for example, raised the quality standards of the goods it is prepared to buy on the open market and put into storage as a means of keeping prices high.

Nevertheless, the Brussels agreement of 1988 does seem to indicate some sort of commitment, if not to reform, then at least to 'capping the CAP' by imposing limits on aggregate levels of farm spending. In the long term, there are new international pressures building for a liberalization of farm policy. In its opening statement to the current round of negotiations on the General Agreement on Tariffs and Trade (GATT) which we mentioned in chapter 1, the United States has called for the abolition of all farm support spending, apart from direct income payments. These international pressures mean that price support policy looks certain to change over the long term, becoming much more market orientated than at any time since the war. With this shift in prospect, reforms in the area of structural policy begin to look more likely.

Policy makers are already aware of the need to manage some of the farming changes that reduced farm spending will bring about. An immediate spur for action here is the threat of renationalization of the CAP if no action is taken and member states proceed to introduce their own schemes of assistance to affected farmers. The next few years should see the development of a stronger EC policy for agricultural structures. The European Commission itself, and to an extent the British Ministry of Agriculture, seem ready to use socio-structural schemes as tools to speed up land use changes in some areas (using the extensification scheme) and slow it down in others (as in the Evironmentally Sensitive Areas). It will be in terms of this management of farming change that environmental interests will find it easiest to engage with future agricultural policy.

The countryside effects

Effective policy reforms like these must necessarily leave their mark on the countryside and its communities, as well as on farmers and their businesses. The temptation for policy makers, absorbed in seeking solutions to budgetary crises, is to ignore these wider effects. A policy of lower prices or quotas, for instance, can be expected to force farmers to manage their land in a number of different and sometimes even novel ways. Rapid and abrupt policy changes, if they were to happen, would give rise to some very dramatic changes in wild places and landscapes as land values plummet and certain vulnerable farmers are driven out of business. In the uplands for example, where, other things being equal, land prices would fall furthest in response to a large withdrawal of public funds from agriculture, it is likely that land released by farmers would be brought up by private forestry companies anxious to afforest large areas of moorland and heathland for financial gain, a point we return to in chapter 3.

More moderate policy changes, phased in to allow farmers time to adjust and therefore resulting in fewer shifts in land use, would still mean changes connected with the introduction of new enterprises and the efforts of farmers to cut costs. The laying off of farm workers is likely to be one of the economies which a farmer with a hired work force will sooner or later have to contemplate. Full-time jobs are expected to be more vulnerable than part-time or casual labour. Economising in the use of fertilizers, sprays and other bought-in inputs, as well as reduced investment in farm machinery and equipment, will mean less business for the large agricultural support industry, both makers and suppliers. Less production will also affect the industries which process and distribute food. The evidence from the imposition of milk quotas is that these 'downstream' and 'upstream' industries will be forced to reduce their labour force once agriculture begins to tighten its belt. A research project funded by the Department of the Environment to look at the countryside effects of CAP reforms has forecast some significant reductions in rural employment for a number of scenarios. Figure 2.16 gives some orders of magnitude of these losses.

Clearly not all parts of the country will be affected in the same way. Many experts are predicting that most change will take place in what they term 'middle countryside', which contains medium quality land sandwiched between good arable land in the South and East and the marginal hill and uplands in the West and North. They argue that even with quite drastic cuts in price support, specialist arable producers in the arable heartland areas will still be able to earn reasonable profits, with little need to diversify into new enterprises or cut costs any further than they have already.

It is unlikely that special payments to farmers in the 'less-favoured' hills and uplands would be discontinued. Existing livestock production in these areas will also therefore continue much as before, at least for the medium

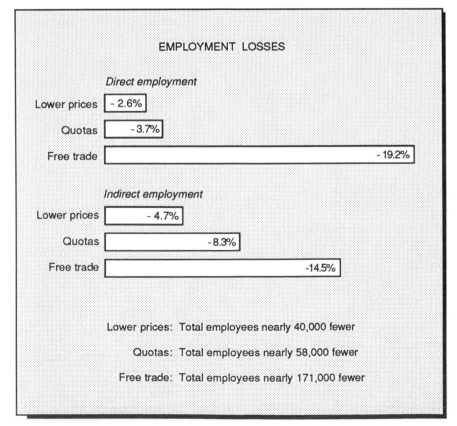

Figure 2.16 CAP change and its employment effects. These three policy options lead to losses in jobs in agriculture and its related industries. Source: Centre for Agricultural Strategy

term, although the increased production of sheep meat from lowland farms for a declining market could have a knock-on effect. However, the weight of evidence seems to suggest that it is within middle countryside that farmers would have to make the biggest adjustments, bringing land out of cereals and putting it down to grass, diversifying into farm tourism and other non-agricultural enterprises and generally streamlining their businesses in line with new market realities. Again, we return to these potential changes in chapter 5.

Reducing nitrogen usage on farms is likely to have a significant ecological impact on the countryside. Current levels of use, for example, are causing severe pollution problems in groundwater that have been linked with stomach cancer and the 'blue baby' syndrome. Recent experiments by the Ministry of Agriculture, Fisheries and Food and the Department of the Environment together, have indicated that the treatment costs for purging water of nitrates to an acceptable level would be roughly the same as a compensation payment to farmers not to use nitrogen at all. This has led to a call in Whitehall for 'low-nitrate zones' in the more intensively farmed parts of the country.

A programme of set-aside and land diversion, operating by itself or in association with price cuts, is likely to have an impact on rural communities and the physical landscape in a number of ways, not all of them welcome. Environmentally speaking, much depends on the alternative uses to which the land is put. Linked to a programme of creative conservation, set-aside could do much to reconstruct landscapes and restore specialized habitats in intensively farmed parts of the lowlands. However, simply abandoning land and allowing it to 'revert back to nature' might be regarded with approval by some ecologists who would like to see more 'wilderness' in the British countryside, although it is feared by others, including many farmers and their representatives. They resolutely reject this option, insisting that diverted land must be put to some alternative use. Forestry is the favoured option and there has been much talk of large scale forestry 'moving down the hill' from the

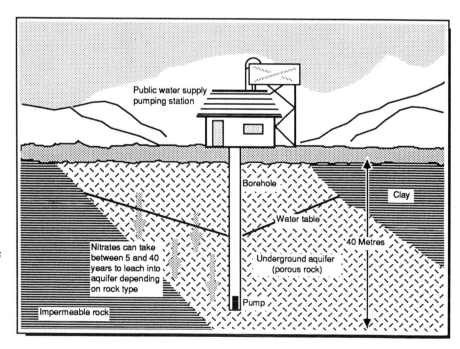

Figure 2.17 Nitrogenous fertilisers in the water supply. Impacts on the countryside can be more than just visual!
Source: J. Blunden and N. Curry

Figure 2.18 The future of farming? Horse riding jockeying for position in the farm enterprise.
Source: *Farmers' Weekly*

unproductive marginal land it has previously been forced to occupy – a development which is unlikely to be greeted with unreserved enthusiasm by those who criticize the damage done by blanket afforestation of open habitat in upland countryside. The diversion of land now in arable production into approved conservation uses, including woodland at an appropriate scale and of an appropriate level, could nevertheless be environmentally beneficial. It is to such considerations and a wider discussion of forestry that we turn in the next chapter.

Chapter 3

Expanding forestry and woodland: barking up the wrong tree?

The historical fortunes of forest policy

From chapter 2, it is clear that one of the principal alternatives to agricultural land use is that of forestry and woodland. This is seen as attractive because in Britain we are unlikely to over-produce timber; indeed we have imported more than 90 per cent of our requirements right through the century. But even if we have no over-production problems, the economic base of forestry is still a fragile one. This is perhaps not surprising when you consider how long it takes to get an economic return on timber, from planting to harvesting.

The balance between our heavy import bills and the economic uncertainty surrounding forestry has led to British forestry policy fluctuating considerably in this century. Since it is instructive to show that such fluctuations are not new, we begin this chapter by sketching in something of the history of our past use of forests and policies towards them. From such a perspective, we move on to examine the economic fragility of forestry today.

The Romans came to an island which was extensively covered with trees, especially with oak, ash and alder on the heavier soils in the lowlands. The invaders cleared some woods for cultivation and carved out tracks for their network of roads. The Angles, Saxons and other later arrivals from the east still faced virgin country for the most part. They penetrated along the Roman roads and up the river valleys – the valley lands were fertile and vital water supplies were available. Tree clearance extended the areas of arable and pasture lands around the villages. Destroyers of woodlands were not given free rein, however. The Saxon kings sought to preserve extensive hunting grounds and the importance of woodlands to local communities for a supply of domestic fuel, construction timber and many other purposes, including beech mast and acorns (or 'pannage') for pigs, was recognized. Severe penalties would be imposed if trees were wantonly destroyed. This conscious attempt to arrest the reduction in forestry area represents one of the earliest examples of nationally-determined rural policy.

William the Conqueror was said to have had a great love for deer, and the Normans, in another phase of forest policy development, designated large tracks of land as royal forests. In those days, a forest was simply a district reserved to the monarch for hunting purposes and the presence of trees was almost incidental. The designation of a forest nevertheless provided protection for the trees and woods within it. In many areas, only essential fuelwood was allowed to be cut and that 'only in the view of the king's forester'. Felling licences have a very long history as a policy instrument!

As the centuries advanced, enclosure movements, developments in agricultural practice and technology, a growing population to feed and house, a decline in the importance of the royal forests and general economic pressures, all combined to produce a substantial clearance of woods. The climax of the agricultural revolution and the final enclosures of the late eighteenth and early nineteenth centuries saw the establishment of a landscape pattern which, in terms of the basic mix of farmland and woodland, has survived to this day.

It was in the fifteenth and sixteenth centuries that the growing need for oak for the navy resulted in widespread acceptance of the notion of planting trees for timber production. Until that time, careful management of natural forests under a coppice system had enabled a steady supply of produce to be harvested to meet local needs. But from the sixteenth century we find ourselves locked into a forest policy cycle that was to last some 300 years.

Figure 3.1 John Evelyn's Sylva. An early example of forest policies designed to ensure a good supply of naval timber.
Source: Scolar Press

The outbreak of a war would find the navy short of ships. Oakwoods would be plundered and the timber immediately converted into 'wooden walls'. A crash programme of planting to produce supplies against a future emergency would follow. At the end of the war, the naval vessels remaining would mostly be neglected – partly out of complacency and partly because of economic pressures in the wake of a conflict which had consumed much of the nation's resources. Many of the young oak plantations would also be neglected as the urgency of war became less dominant in people's minds. Ships that had been mothballed would tend to rot because of the unseasoned timber used in their construction. Thus, the next outbreak of war found the nation as unprepared as the previous time and the cycle of crisis action followed by neglect would be repeated.

The recurring demand for oak for shipbuilding came to an abrupt end with the industrial revolution of the nineteenth century. Iron and steel superseded timber not only in the navy but also for a wide range of industrial and domestic purposes. Moreover, coal had substantially replaced wood as fuel. The most important timber requirement was for softwood, the wood of coniferous trees, which could be much more easily worked by the new machinery than the traditional oak so lately prized for its hardness and durability. Softwoods could also be imported in abundance. The future of the traditional British hardwoods looked bleak and very many woodlands fell into neglect. New powered machinery made tree clearance easier too, and as

Figure 3.2 Typical eighteenth-century parkland – Chatsworth. These landscapes designed to provide status and pleasure for the original owners, are today a source of enjoyment for the many. *Source*: Countryside Commission

agriculture prospered, the woodland area was further reduced. This nineteenth-century phase of forest policy was therefore characterized by decline.

Yet some of the old oakwoods which are highly valued by conservationists today originate from plantings in the wake of the Napoleonic Wars. And many of our most significant parkland landscapes were laid out in the late eighteenth century when agricultural prosperity was such that landowners could afford the luxury of parks and grand gardens. What landscape features are we now establishing which will endure for the benefit of the twenty-second century?

By the early years of this century, we in Britain had a lower proportion of our land covered with trees than at any other period in its entire history – about 5 per cent. Around 90 per cent of our timber requirements were imported, mostly softwoods from Scandinavia and North America. Many of our woodlands were in poor condition and the oaks prompted by the Napoleonic Wars were semi-mature. In 1914 we were launched into the same cyclical pattern of crisis action as had been observable over many previous centuries.

The First World War saw us thrust back on home timber production, as imports were cut off. Any trees of usable size were liable to be felled and about one-sixth of all our woodlands were devastated. The strategic importance of boosting home production was convincingly demonstrated and in 1919 the Forestry Commission was set up, to establish state plantations, to administer policy for private woods and to develop a more coherent national policy for timber.

The Forestry Commission's main policy objective at that time was to build up a reserve of timber against the possibility of another major war. Minor objectives were to provide some insurance against general world shortages of timber and to help reduce the drift to the towns by providing new jobs in the countryside. Thetford Forest in East Anglia, for instance, was mostly originally planted in the late 1920s and early 1930s as a job creation scheme.

As memories of the war faded and new economic and political problems mounted, so much of the urgency again disappeared from forestry policy. The Commission only narrowly survived a proposal to abolish it and its grant-in-aid was reduced. From 1919 to 1939 the Commission did not achieve its planting target while, in the private sector, plantings barely kept pace with fellings. The home grown timber trade remained in the doldrums and woodland neglect was widespread.

The crisis cycle began all over again in 1939. Private woods were again devastated and a new and larger official target was announced – two million hectares of productive woodlands in Britain by the end of the century. It was estimated that such an area would yield about one-third of our annual consumption (at pre-1939 levels) and would enable us to be self-sufficient for at least three years in the event of yet another emergency. Shortly after 1945, assistance to private owners to manage their woodlands and establish new ones was greatly expanded and the Forestry Commission's planting averaged over 24,000 hectares per annum at its peak. Strict controls were maintained over tree felling in order to conserve timber.

This historical cycling of forest policy has not abated even since the Second World War. By the late 1950s, the strategic argument for forestry had once again weakened, and the decline of hill farming led to more emphasis being placed upon the role of forestry in providing employment and other social and economic benefits. The downturn in forestry support was compounded at the time by a closer examination of the economic rationale of forestry. Important question marks were raised against the projected low returns on

Figure 3.3 Afforestation's artful aid. Amenity and job creation roles of forestry have long been recognized.
Source: Punch

public investment in upland plantations, but estimates that employment in forestry could be double that in hill farming over the same land area proved irresistible.

Through the 1960s emphasis continued to be placed on the economic and social development role of forestry, especially in remote rural areas. Efficient timber production remained the Commission's main objective and some of its own early plantations were beginning to come into production. Two other important aspects of forestry assumed much greater prominence in forest policy – the effect of plantations on the landscape and demand for public access to forests for recreation. Many plantations established in the 1930s and 1940s had emerged as dark, geometric shapes out of harmony with the landscape – and these unattractive masses were virtually impenetrable for the purposes of recreation. The Commission began to give much more attention to landscaping and to make more positive provision for public recreation as a central part of policy.

Meantime, the private sector had been slowly gaining confidence nurtured by policies that provided grants and tax concessions. Private organizations were engaging in upland conifer planting, on behalf of investors, on a scale comparable with Forestry Commission activities. Many lowland woods were once more brought under active management, with broadleaved species often giving way to conifers and prospects for the home-grown timber industry began to improve. By the early 1970s private woodland owners were planting well over 20,000 hectares per annum, about as much as the Forestry Commission and in the peak year of 1971, private and public sectors combined to plant over 51,000 hectares.

Inevitably, this expansionary phase of forest policy was not to last. A governmental review of this policy in 1972 again highlighted the very low rate of return in public investment and a reduction in the Commission's planting programme was announced. The principal grant scheme for private woodlands was suspended. The introduction of capital transfer tax in 1975

Figure 3.4 Contrasts in forestry. Regimented conifers have given way in the 1980s to a less geometric approach, using deciduous trees (see figure 1.5) and more closely resembling our woodland heritage. *Source*: Countryside Commission

further eroded confidence. By 1978 private planting had slumped to less than half the annual rates at the start of the decade. This time the 'crisis' trigger was not a military threat from abroad, but economic and political changes at home. In the 1980s, as we shall see, the private sector made a recovery while the Commission's planting activity has remained far below the level of the early 1970s. What has increased, however, is the scale of the Commission's harvesting and marketing operations.

Present patterns of forestry

The combined efforts of the Forestry Commission and private woodland owners have doubled the wooded area of Britain this century to around 10 per cent of the country. Our proportion of woodland is still lower than most other European countries: only Ireland is substantially lower as we can see in figure 3.5.

Given the frequent changes of national policy, it is remarkable that the target of 2 million hectares of productive woodland by the end of the twentieth century, set in 1943, has just been achieved. Around two-thirds of this productive woodland is coniferous and the dominant species is the Sitka Spruce which was introduced to this country from the west coast of North America. It has been very widely used on upland sites on the wetter, western side of Britain. The extent of its use is an indication, not that acid bogs and windswept hills are the ideal conditions for it to flourish, but rather that

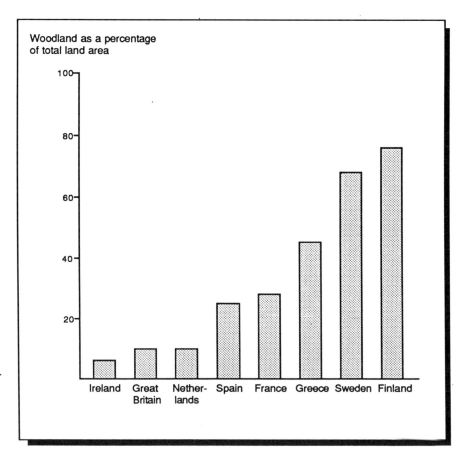

Figure 3.5 Europe under woodlands. Britain is near the bottom of the league in the proportion of its land under timber.
Source: D. Nicholls

almost no other species has been found even to survive such conditions! However, should we have planted such areas at all?

Overall today, the area of private woodlands is rather higher than that of state forests (about 55 per cent of the total, as against 45 per cent) as we indicate in figure 3.6. The composition of the two sectors differs markedly, with most of the area of broadleaves in private ownership. Broadleaves, including small coppice areas, comprise 46 per cent of private woods compared with a mere 6 per cent of the Forestry Commission's total. This is a reflection of several factors. As we have already seen, the history of much of the private sector, particularly the lowland woodlands, has its origins far back beyond the founding of the Commission and the widespread use of introduced conifers. The higher fertility of many of the private lowland sites means they are far better suited to broadleaves than is exposed moorland. The wider range of management objectives of many private woodlands, requires a diversity of species and a large broadleaved component is often considered essential for the reasons of aesthetic satisfaction.

We still import nearly 90 per cent of our timber requirements, currently costing around £6 billion a year. Home production is now increasing steadily and could double over the next fifteen years. But will demand go on increasing as living standards rise? Paper accounts for a third of our imports.

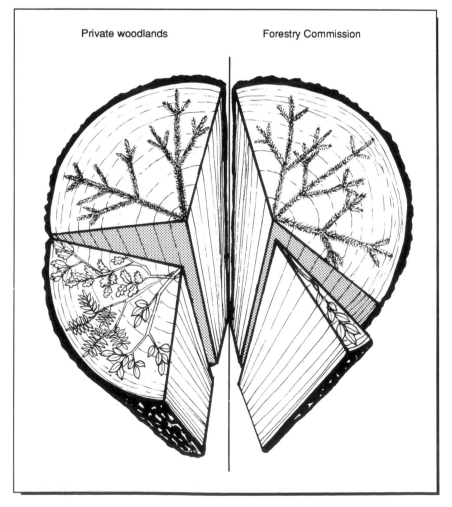

Private woodlands   Forestry Commission

Figure 3.6 Private and public forestry.
Although ownership is roughly equally divided between two sectors, most of the broadleaved woodlands are in private hands.
Source: D. Nicholls

Some may argue that the electronic revolution will drastically reduce the demand for newspapers and books and that we will no longer write letters. Although we may no longer write to each other to the extent that we did at the end of the last war, there is evidence to show that television, together with rising standards of education and new marketing methods for books, have vastly expanded book production over the last decade. At the same time, a new generation of users of micro-processors at home and in the office has led to the development of a huge market for computer paper. Meantime, 'junk mail' pours through our letter boxes in ever increasing quantities. Wherever the truth of the relationship between the supply and demand for paper really lies, the forestry issue is additionally complicated by the pressures to expand the area under trees as part of set-aside, as we discussed in the previous chapter. But again this does not necessarily bear any relation to the demand for timber or the economic worth of forestry.

Organizationally, the relationship between the public and private forestry sectors seems to have worked quite well, as far as the foresters are concerned. The Forestry Commission has provided a great deal of advice to the private sector – until recently, largely free of charge – on the basis of growing experience and research. One of the most important relationships between the sectors in many parts of the country is the lead which the Commission gives in marketing and attracting wood processing industries. As the biggest producer in most regions, the Commission is able to enter into supply contracts to provide stability for the processors and thus enabling smaller-scale producers to have an outlet for their timber which could not be sustained by the private sector alone.

It is also the Forestry Commission which administers the various grant schemes for private owners and among foresters is generally perceived as doing so in a reasonably sensible and sympathetic manner. In 1986–87 grants to the private sector account for something like £7m out of a total Commission budget of over £150m per annum. (The beneficial tax arrangements, which we shall discuss later, have of course been dealt with by the Inland Revenue, not by the Forestry Commission). However, these public expenditures on forestry do give an indication of its economic marginality, but do bear in mind, as we noted in chapter 2, that total support for UK farming was £1800 million.

Criticism of the relationship between the public and private sectors in forestry focuses particularly on the way in which all the forestry interests have been largely free to go their own way. Tree planting is not subject to planning permission, although, consultation with National Park planning authorities is now demanded in such areas. Tree felling is subject only to Tree Preservation Orders or permission from the Forestry Commission where grant-aid is involved. Acceptance of a private proposal for a grant-aided scheme or a major Commission development also involves consultation with various public and private bodies. There is a widespread view, however, that forestry interests need to become more publicly accountable than is suggested by these measures and subject to wider control and influence.

Economic fragility and forest investments

Every few years, the public investment in forestry that we have just discussed comes under the spotlight of economist and accountants. Public investment in forestry, after all, diverts resources from other projects and it is reasonable to examine the returns yielded for the nation. Judged by the normal criteria for investments, the performance of forestry is poor. The review of the early 1970s which precipitated a sharp fall in forestry activities, included a cost benefit analysis which indicated that, even including the social benefits such as employment and public recreation, the return on the national forestry investment was only within the range of 3 to 5 per cent – whereas the Treasury test rate at the time was 10 per cent.

Even though the target rate for public sector trading operations is now 5 per cent, the Forestry Commission is set a target real rate of return of just 3 per cent per annum on the total forestry estate. That target is only achieved with the aid of subsidies for non-commercial activities. The Commission expects to achieve return of only 2.5 per cent on investments in new planting in 1987–90. On the other hand, in more favourable planting areas, financial returns of over 5 per cent are sometimes achieved.

Conventional investment analysis thus points to the conclusion that investment in forestry shows an unacceptably low rate of return; that both public and private investors would do well to put their money elsewhere. So why don't they?

It must be said that not all economists readily accept the analysis and conclusions of such cost–benefit studies. There are two broad groups of reasons for that. On the one hand, there are fairly major methodological problems which cannot be easily overcome. For example, what assumptions should be made in relation to timber prices fifty or more years hence? How should the cost of creating new jobs in forestry be measured? How do you put a figure on the recreational benefits of a forest which is visited by hundreds of thousands of people each year for periods varying from a few minutes to several hours at a time? On the other hand it is argued that the scope of the analysis is too narrow, that the whole wood processing industry and allied services which are dependent upon British forest produce should be taken into account. Thus, for example, the net subsidy per job in the industry as a whole is much less than if only those employed within the forest are considered. It is estimated that some 40,000 people are employed in Great Britain as a result of home-grown timber activities.

Typically, government response to such analysis is ambiguous. In 1972, a White Paper accepted the conclusions of a return of around 3–5 per cent compared with the target of 10 per cent, yet declared that further investment would be undertaken in the interests of employment in remote areas. Would it not have been more logical to have said that the analysis seriously under-valued the significance of new jobs and, in fact, the true return was greater and justified further public expenditure? The pursuit of the rural employment objective has meant that much of the Forestry Commission's recent investment is expected to yield (through timber production) only around 1.25 per cent. The House of Commons Committee of Public Accounts in 1987 expressed concern that full regard should be paid to the comparative costs of such job creation, as well as the benefits, and to the additional opportunities for extending the Commission's profitable activities in tourism and recreation. The case for forestry in terms of helping our balance of payments was also felt to be unproven. This House of Commons Committee found that 'across many of the Commission's activities there was insufficient assurance on the extent and quantification of the benefits achieved or how far these were commensurate with the resources used to achieve them'. In short, we do not really know how much our forestry is costing or what we are getting out of it!

In 1980, in line with the government's general view of privatization, the Forestry Commission was directed to sell off a slice of its assets, with a view to reducing the annual sums voted by Parliament. Sales included both small, detached forests and high quality compact blocks, together with some of the Commission's dwindling reserve of plantable land. In seven years £100

Figure 3.7 Saw-mills in Scotland. Timber processing in plants like this one at Kirkoswald in Ayrshire, create many more jobs than the forest itself.
Source: Forestry Commission

million worth of assets had been sold, but it is unlikely to have had a significant effect on the overall rate of return on public forestry investment.

Private forestry investment appears to show returns which are similar to the Commission's figures for comparable sites. Private investors, however, usually have financial help from the state, as we shall see later, and the net rates of return may be significantly higher and attractive to wealthy individuals and, in some cases, to financial institutions. The latter are particularly attracted by the prospect of tax-free capital gains which can then be re-invested to yield a steady annual income.

So forestry investment will always be marginal in an economic sense, but there could still be arguments for expansion based on uncertainties about our requirements for timber or timber products in a hundred years from now and of our ability to buy all we need on the world market in the future. Many would see an important insurance element in forest investment, especially in the higher value hardwoods.

Tropical hardwood forests are declining at a rate of several million hectares a year. We could boost our hardwood production so as eventually to reduce our demands on tropical forests the destruction of which seriously poses world ecological problems. If we criticize other countries for large scale forest clearance, we should remember the extent of our own clearances over the last two thousand years and possibly use this as a justification for continuing a positive re-stocking programme.

Future investment by the Forestry Commission is clearly a matter of policy of successive governments. The expansion of the Commission's forests has slowed right down, but it seems likely that they will at least maintain their existing estate – unless a further substantial proportion is to be privatized. The cumulative nature of forest investment may continue to attract certain types of financial institution, particularly if investing in agriculture where policies are aimed at curbing over-production remains unattractive for some years to come. Above all, the traditional estate owners, having clung to their woodlands through deep depressions, are not likely now to turn their backs on those investments. Pressures on landowners arising from declining returns to agriculture, however, may decrease the prospects for significant woodland expansion. It is paradoxical that just when farmers are being asked to use less land for food production they can often least afford to switch from food production into forestry. The many benefits traditionally derived from estate forestry represent much of positive value to the nation as well as to the individual and it is inconceivable that public subsidies will not be maintained.

Assisting the private sector

From the beginnings of the Forestry Commission in 1919 it has been accepted policy that the Commission should provide financial assistance to encourage the planting and good management of private woodlands. The first grant schemes related to scrub clearance and other ground preparation, as well as planting, and larger grants were payable to corporate bodies rather than to private individuals. In 1927, planting grants were fixed at £5 per hectare for conifers and £8 per hectare for hardwoods, irrespective of ownership.

It was not, however, until after the Second World War that the private sector received assistance on a scale which evoked a very positive response. Under the Forestry Act of 1947, the key feature of future grants for private woodlands was the Dedication Scheme. The planting grant (for all species) became £25 per hectare and the following year an annual management grant was introduced. For woods still aided under this original Dedication Scheme, the planting grant is now £110 per hectare, with an annual management grant of between £2 and £4.80 per hectare.

Two vital conditions were attached to acceptance for the Dedication Scheme. Firstly, landowners had to enter into a dedication covenant, binding them and their successors to maintain the use of the land as woodland in

Figure 3.8 Woodlands on the farm. The planting of trees is one form of diversification that farmers are being encouraged to undertake. But can this be more than a marginal economic pursuit? *Source*: Countryside Commission

perpetuity. Secondly, the then owner had to agree to manage the woodlands in accordance with a plan of operations approved by the Forestry Commission. Such plans were regularly revised and covered all aspects of establishment and management of plantations. One important bonus for owners was that felling operations in accordance with the approved plan were deemed to be covered by a felling licence.

Various other grant schemes were introduced, sometimes only for a few years, but it was the Dedication Scheme which laid the foundation for confidence within the private sector and which set the scene for major expansion of private forest activity. There was thus a shattering blow to that confidence when, without warning, the Dedication Scheme was suspended in 1972. The introduction of a new scheme in 1973, with higher planting grants, especially for broadleaves, did not immediately repair the damage. Several other changes to forestry grants were made during the 1970s and then a new scheme of grant aid was introduced in 1981.

The 1981 Forestry Grant Scheme, for simplicity of administration, took the form of a planting grant only but, compared with assistance still payable on woods remaining under one of the old schemes, the grants were generous. The minimum grant, payable for plantations of conifers exceeding 10 hectares, was £240 per hectare. Smaller areas received proportionately larger grants, in recognition of the higher unit costs of fencing and most other operations. Broadleaved plantations attracted an extra £230 per hectare. Eighty per cent of the grant was payable on planting, with the balance paid after five years, subject to satisfactory establishment. Forestry Commission approval was required, but Commission staff were not involved in the preparation of detailed management plans as under the old Dedication Scheme. One novel condition of the grant was that the owner had to be prepared to discuss the question of public access with the local authority, but there was no binding commitment to allow access.

In 1985, the Broadleaved Woodland Grant Scheme was introduced. This scheme was designed to encourage the rehabilitation of old broadleaved woods, by planting or by natural regeneration, as well as the establishment of new broadleaved plantations. Whereas timber production had to be the primary objective for grant aid under the Forestry Grant Scheme, the Broadleaved Scheme was intended to encourage a range of management objectives, of which timber production had to be one, but not necessarily the most important. Again, landowners were required to discuss possibilities of public access, but without obligation. The rates of grant were significantly higher than for broadleaves under the Forestry Grant Scheme, ranging from £600 per hectare for large areas to twice that in the case of woods under one hectare. These broadleaved grants were perhaps particularly helpful to the small-scale woodland owner and the initial response to the scheme was encouraging.

The forestry grant scheme has become further complicated by the new Farm Woodland Scheme. In an official attempt to induce farmers to reduce the area of land used for food production (and hopefully reduce undesired food surpluses) special grants are to be paid for certain newly established plantations. Under this scheme, a new plantation could attract a grant of up to £190 per hectare *each year* for 20 to 40 years. Annual grants on that scale are a major departure from previous forestry grants and represent recognition that the farmer's real need is not just for an alternative crop, but for an alternative income. The initial scheme is limited to three years and a total of 36,000 hectares. Undoubtedly there will be at least 12,000 hectares a year coming forward for planting under the Farm Woodland Scheme. How much of that might have been planted anyway is an interesting question for

Figure 3.9 The timber harvest. Under the Dedicated Woodland Scheme, felling did not require a separate licence.
Source: Countryside Commission

Figure 3.10 Broadleaved woodlands in the Lake District. The Broadleaved Woodlands Grants Scheme is encouraging planting to create landscapes like this for generations to come.
Source: Countryside Commission

speculation. The contribution of that scale of afforestation to the reduction of agricultural surpluses as a process of agricultural set-aside, however, will be minuscule!

Grants are only part of the story to aid private forestry. Tax concessions have provided far more important incentives to plant trees on a large scale. There can be little doubt that the area of private woodlands would not have shown substantial increase in the last 30 or 40 years in the absence of a favourable tax system. There was, however, mounting criticism of the tax regime which provided the greatest incentives for large-scale coniferous monoculture and the greatest net benefits to those with the highest incomes. The criticism came, on the one hand, from individuals and organizations (including, for instance, the Nature Conservancy Council) concerned at the environmental impact of large-scale conifer planting – the impact on wildlife, on the landscape, on opportunities for outdoor recreation pursuits, and so forth – and, on the other hand, from those concerned at the apparent inequity of offering far greater help to exceedingly wealthy individuals seeking a 'tax haven' without going abroad, than to traditional woodland owners and farmers who were struggling to maintain their woodlands in the face of rising costs and (over long periods) declining real timber prices.

At last, in the 1988 Budget, the Chancellor of the Exchequer bowed to the criticism and effectively ended the main tax concession by removing commercial forestry from income tax relief altogether. It is, however, worth summarizing the former position so that we may later attempt to assess the impact of the tax changes.

In essence, it had been possible to arrange for all the costs of establishing a plantation to be offset against taxable income received by an individual from other sources. Thus, if the new forest owner were taxed at a top marginal rate of, say, 60 per cent, the net cost of the plantation would be only 40 per cent of the nominal cost as a result of the reduction in his tax bill. That provision, it should be stressed, was no different from the treatment of losses under Schedule D in relation to any other kind of business. What distinguished forestry for the purposes of income tax was the possibility of enjoying the proceeds of the timber harvest virtually tax free, under Schedule B. To gain the benefits of a tax-free harvest income after enjoying tax relief under Schedule D necessitated a change of ownership.

There were, however, many devices for achieving that without parting with effective managerial control over the woods. In any case, many of the absentee investors were only too happy to sell, and the ownership condition was only a significant obstacle for some owners seeking to maintain the continuity of a family tradition. Very often a transfer would occur on the death of the owner who established the plantation and long before it was ready for final harvesting. These income tax provisions were obviously worth most to the highest taxpayers. Landowners who paid little or no tax derived little or no benefit. Quick growing conifers for clear felling seemed the way to maximize the tax benefit.

As with previous kinds of death duties, there are special inheritance tax provisions favourable to forestry, notably the ability to defer payment of the tax until standing timber is felled or sold. Tax is then payable on the net proceeds of sale which, following growth and maturity of the trees over several years, may be far higher than the value as at the date of death. The opportunity to defer may be extremely valuable, nevertheless, in easing cash flow problems arising from the need to pay the tax on other property shortly after the death. Conversely, trees represent a potentially valuable store of capital which may be realized to meet tax or other exceptional liabilities. Other tax concessions are reinforced by the exclusion of growing timber from capital gains taxation.

The combined effect of tax concessions could increase the rate of return on woodland investment from, say, 3 per cent to 7 per cent or more, depending on the individual's circumstances. Even after the 1988 Budget, tax concessions other than those for income tax serve to increase this rate of return well above 3 per cent. But how important are the various forms of state aid to private forestry? Where would we have been without them and what is needed for the future?

Certainly, the importance of the income tax provisions is not in doubt. Very little new planting would have taken place without them, other than on established forestry estates or where the primary motivation was other than commercial timber production. A great deal of planting has occurred in connection with land management for sporting purposes and, in the last twenty years, very many small areas and awkward field corners have been planted up for a variety of reasons to do with 'amenity' and conservation, but, in terms of aggregate area, new planting has been overwhelmingly in upland conifer plantations.

Research at Cambridge University, spanning the 1960s and mid-1980s, has concluded that the recent impact of grants and taxation on the management of lowland woodlands has been fairly marginal for most landowners. If an owner were inclined towards positive woodland management, whether for reasons of estate profitability from timber production, for one or more indirect benefits (from shelter to shooting) or out of sheer love of trees, then the chances are that he or she would go ahead with the management, taking advantage of any fiscal incentives as a bonus to be picked up along the way. Tax and grant benefits were seen as encouragement, but were far less likely to be seen as a critical determinant of management policy.

The 1947 Dedication Scheme and the income tax regime, which was popularized by the Economic Forestry Group in the 1950s, together gave

Figure 3.11 Amenity on the farm. Small planted areas such as field corners have done much to enhance the amenity of the agricultural landscape.
Source: Countryside Commission

woodland owners confidence for investment. In the immediate post-1945 era, grants were probably more important than tax concessions, but that soon changed. No one (outside the Inland Revenue at least) knows exactly the value of the income tax concessions, but at the time of their abolition in 1988, the most conservative estimates talked of £10 million and some estimates suggested a figure in excess of £20 million per annum – that is, three times the annual value of forestry grants, but still only just over 1 per cent of £1800 million we spend in Britain each year supporting agriculture.

In the wake of the removal of commercial forestry from income tax and corporation tax, yet another new grant scheme was announced in 1988 which is apparently intended to support what the government termed 'the sensitive yet vigorous expansion of forestry'. A re-statement of an annual planting target of 33,000 hectares of new trees suggested that the 1988 Woodland Grant Scheme was intended as a direct replacement for the lost tax incentives. The new levels of planting grant, range, for conifers, from a minimum of £615 per hectare for areas over 10 hectares to £1005 per hectare for small areas (less than 1 hectare) and, for broadleaves, from £975 to £1575 per hectare on the same basis. Seventy per cent of the new grant will be payable at the time of establishment; 20 per cent after 5 years and the remaining 10 per cent 10 years after planting, subject to satisfactory management. If planting is undertaken under the Farm Woodland Scheme, the 1988 rates of planting grant are payable for broadleaves, but the earlier and much lower rates will apply to conifers.

Will the new grant schemes result in the achievement of the overall planting target 33,000 hectares? Will the Woodland Grant Scheme be as powerful an incentive as the former tax concessions? Even more to the point, will the new scheme achieve not only timber production objectives, but also sensible environmental objectives?

Forestry being such a long-term investment with a very low conventional return, it seems necessary for state aid to private woodland owners to continue in some form, if we wish to have well-managed private woods. We shall return to appropriate objectives and forms of assistance in the final part of this chapter.

New planning and management initiatives

For more than half a century, British forest policy has been dominated by a desire to produce larger quantities of mature timber so as to reduce our dependence on supplies from abroad. Large-scale conifer afforestation seemed to offer the quickest and cheapest route to that goal. Private and public sectors alike were thus mobilized to that end and, it must be acknowledged, responded with a large measure of success. It has become fashionable in some quarters to blame the forestry industry for 'doing the wrong things' just as farmers are widely criticized for excessive cereal production. The fact is, however, that we have broadly achieved the pattern of rural land use and management that we, as a nation, have been asking for. If 'blame' is to be apportioned, we should perhaps not entirely exclude ourselves!

Now that there is broad agreement on the need for a different emphasis in forestry away simply from timber production, what are the possibilities? 'Amenity' plantations, tree cover for game and public recreation sites have already been mentioned. It is fashionable to talk of multi-purpose woodland use, almost as though the concept was a discovery of ecologists or planners of the late twentieth century. How easily we forget the diversity of benefits derived from medieval woods. The coppice-with-standards system of management yielded everything from walking sticks to house timbers, provided shelter for man and beast, fodder for domestic and game animals,

Figure 3.12 Coppicing in woodland management. Such a system represents a return to more traditional systems of forest husbandry. *Source*: Countryside Commission

sporting opportunities, dominant landscape features and a rich variety of habitats for flora and fauna. Such woodlands were managed by people who had an appreciation of a delicate ecological balance, though they lacked our modern scientific knowledge. Sound woodland management must be based on proper ecological principles and, after a period of pursuing an 'unnatural' style of forestry we are, perhaps, returning to approaches which are more in tune with both the laws of nature and the needs of modern society.

In 1987, the Countryside Commission published a major statement *Forestry in the Countryside* and suggested that national forestry policy should be based on multiple objectives, namely, to produce a national supply of timber as a raw material and as a source of energy; offer an alternative to agricultural use of land; contribute to rural employment either in timber industries or through associated recreation developments; create attractive sites for public enjoyment; enhance the natural beauty of the countryside and create wildlife habitats.

Somewhat piously, the statement adds: 'In future, all forestry proposals should aim to fulfil in different measures all of these objectives'. Yes, but as the Countryside Commission recognizes, not all present or future forests will be capable of significant contributions under *all* of those headings. We may be in broad agreement on the range of objectives; the crunch comes when we have to sort out priorities.

To be fair, the Countryside Commission has done a great deal to try to practise what it preaches. It offers grants for amenity tree planting; it has been involved in a number of projects aimed at improving agricultural landscapes and it has promoted important pioneering work in relation to the use and appearance of land in urban fringe areas. The Commission is one of the sponsors of Project Silvanus, in south-west England, and of Coed Cymru in Wales. The concern behind these initiatives is to make available to landowners wishing to plant or manage a wood a single source of advice on a whole range of matters from the siting and design of a plantation, and what to plant and how, to the employment implications and the marketing of the expected produce.

Silvanus was the Roman god of farms and the wildwood. The Silvanus Project was launched in 1986 under the sponsorship of the Countryside Commission, the Ministry of Agriculture, Fisheries and Food (MAFF), the Nature Conservancy Council, the Development Commission, the Forestry Commission, the Dartington Hall Trust – with a long record of enlightened rural management and development – and the Department of the Environment (DoE). By offering a wide range of advice to landowners, the project aims to get 25 per cent (10,000 hectares) of all neglected woodland in southwest England under management within ten years. A trading company has been set up with a view to making the project financially self-sufficient within the same period. Conservation, employment, exploitation of woodland produce and supplementing farm incomes are key themes of the project. Where the projects own staff cannot provide all the answers or services desired, they are well placed to co-ordinate other specialists.

The initial reaction to Silvanus and Coed Cymru has been encouraging. Many woods are now facing management for the first time for many years – even decades or centuries. A pre-requisite of integrated, balanced land management is integrated balanced advice and that is more likely to be forthcoming through an agency like Silvanus than through more specialized commercial or public organizations. It is difficult for MAFF officials to get away from thought of 'efficient food production' and for Forestry Commission staff not to be dominated by 'efficient timber production'. To say that is not necessarily to criticize in either case, but merely to stress the need for a different source of advice. On the commercial side we do not expect timber merchants to be experts in protecting fragile habitats!

The Countryside Commission statement also contained proposals for two specific forestry initiatives which are worthy of serious consideration. The first is the creation of several forests in urban fringe areas which in decades and centuries to come would fulfil the sort of roles currently played by say, Epping Forest and Cannock Chase. The attractiveness and recreational value of many areas close to large cities are in desperate need of enhancement. Tree planting and woodland management with landscape and recreation needs uppermost among management objectives would make a significant contribution in the long term.

The second initiative proposed is the establishment of a major new forest in the English Midlands. Here, the model is the New Forest in the south – a remarkable blend of woodland, open heath, grassland, villages and hamlets adding up to an area of scenic and ecological importance and a source of immense pleasure to millions of visitors each year. An area of, say, 40,000 hectares is envisaged, mostly remaining in private ownership, but with inputs from a range of public and voluntary organizations, in co-operation with the landowners, possibly through some form of trust.

The idea of a 'new New Forest' will take years to come to fruition, but it is an example of the kind of bold thinking which is needed if we are to develop a coherent set of rural land use policies for the future. We need more than marginal adjustments to policy aims in the light of changing circumstances. We need clear statements of new priorities for forestry and other land uses – just as we were given every time we were short of timber during a war. We need bold initiatives and a willingness to take risks in departing from conventional calculations of returns to public investment. In the last part of this chapter we return to our starting point – we look now prospectively, at national forest policy and its implementation.

The way ahead – a growing forest?

Forests represent resources of great versatility. They are renewable resources, capable of management on a sustained yield basis. They offer an

Figure 3.13 A new New Forest? The New Forest in Hampshire can provide a model on which to base the development of other amenity forests in Britain.
Source: Countryside Commission

enormous diversity of 'products' – from fuelwood to furniture, from pulp for paper to packing cases, from fence posts to roof trusses, from fibre board to fodder, from picnic sites to protected species, from sport to shelter. Although we in Britain have a minute fraction of the world's 4 billion hectares of forest, we should not forget the global role of photosynthesis and the conversion of carbon dioxide into oxygen. If that role is not maintained, life as we know it will cease.

As a nation, we now have a much greater awareness of the possibilities for forestry and there is no longer the imperative of the rapid establishment of a strategic reserve of timber. At the same time, the traditional constraint on the use of land for forestry, the overriding priority of food production, has been relaxed. Now, perhaps for the first time for centuries, is an opportunity to re-think forest policy in the context of a re-appraisal of policy for the country-side as a whole. It is surely sensible that the future for forestry (as for any other land use) be worked out in a context which considers the wide range of other uses to which our countryside may be put. What we want from our countryside in the twenty-second century and beyond and how we can achieve those goals so as also to ensure maximum flexibility of policy and achievement for the twenty-second century and beyond, are important issues.

The broad objectives for forestry proposed by the Countryside Commis-sion – which have long been espoused by the Forestry Commission too – are a reasonable starting point. We want, and will continue to want, timber (or at least cellulose in some form), rural employment, recreation oppportunities, attractive landscape and wildlife habitats. These forestry goals have to be woven together with other objectives for the countryside. In many cases, goals for different land uses will be complementary: in others there will be sharp conflict.

Perhaps we should establish clear priorities for forestry. Certainly, one of our past failures has been that laudable national goals have not been adequately translated into local policies for implementation. The building blocks of a national strategy might be local rural strategies prepared, perhaps, on a county basis. Cambridgeshire County Council, for example, has produced a rural strategy for the county, embracing the management of the Council's large farms estate as well as concerns of the agricultural community at large, problems of small rural settlements, employment and the rural economy, recreational access to the countryside, wildlife conserva-tion and the planting of trees and hedges in a landscape which has fewer trees and woods than any other English county. The possibility of substantial tree planting to form local amenity forests near the cities of Peterborough and Cambridge is mooted.

Perhaps it is at that kind of scale that broad strategic plans should be prepared and at which the main focus of implementation might lie. Blanket tax concessions and uniform national grant schemes take no account of wide local and regional variations – variations in the needs and demands of local communities and in the practicalities and costs of forest creation and maintenance. What is in the interests of conservation or recreation in one area may not be appropriate in others. Forest management must reflect differences in soil, topography and climate and quite different tree species may be required to achieve similar goals in different locations.

Grants for the private sector need to have much more flexibility than a simple conifer/broadleaf distinction. The emphasis might be on achieving particular targets rather than a rigid insistence that 'broadleaves are always best' to allow for some local flexibility.

National and local objectives need to be as positive and specific as possible. The negative goal of reducing the area of land which is intensively farmed is not an acceptable basis for forest policy. This is despite the prevailing view, as we saw in the previous chapter, that one means of reducing agricultural surpluses is to turn more land over to forestry. The Countryside Commission in developing a co-ordinated national forestry policy has given a number of useful leads. Maybe in targeting the task of bringing existing woodland back under management, we should be allowing the precise objectives to vary from one region to another and from farm to farm. Perhaps a target of, say, 50,000 hectares of urban fringe woodland to be established primarily as recreational resources by the end of the century would be best. Why *not* go for a new forest of 40,000 hectares in the Midlands? At the same time, bearing in mind other rural interests, it might be appropriate that we should be protecting some areas from major change, including tree planting. Surely all forestry activity in Environmentally Sensitive Areas should be subject to much tighter scrutiny – not just a virtual ban on conifers. Should the Caithness 'flow country' not be declared a protected area, regardless of the pattern of afforestation which is contemplated?

In terms of land area, the private and public sectors of British forestry are fairly evenly balanced. This pattern, which has evolved since 1919, is typical in many mixed economies. The 1980s have seen a slight, but deliberate, shift of emphasis in public policy towards the private sector – the privatization of some Forestry Commission assets coupled with an expansion of private planting targets while the Commission's targets have been lowered. The 1990s may see further shifts in that direction, but any major shift is more likely to be motivated by political doctrine than by compelling arguments of resource economics or national welfare. The Forestry Commission has acquired a substantial body of knowledge and skill in growing trees under a wide variety of British conditions. These skills and the ability to implement policies for, say, public recreation provision should allow forestry development to be enhanced by public ownership. If the state cannot take a long term view of investment, who can?

The Forestry Commission, in its forest enterprise role, could be free to undertake some new planting in accordance with local or regional rural strategies, though inevitably it will be primarily concerned with harvesting and re-planting in accordance with those same strategies. The Commission, however, need not be the sole Forest Authority in relation to the private sector. Administration of grant schemes and detailed advice might continue to be the responsibility of the Commission – there is no substitute for silvicultural expertise in such matters. But the broad framework for control could be set within national guidelines by a county or regionally-based authority.

The county authorities need not be orthodox committees of county councils, for it is important that a wide range of interests are fully involved. If the county council has the ultimate responsibility, most of the consultation, debate and policy recommendations could come from a wider forum on which private and public forestry and timber-using interests are represented, together with other bodies which have interests in other types of woodland product, such as recreation, nature conservation and landscape enhancement. The county authority might both prepare the local strategy and oversee its implementation. Major afforestation schemes (say, more than 10 hectares) could require specific permission from the authority, regardless of whether any financial assistance is involved. Some extension of public control over private forestry could be the quid pro quo for continued (and possibly more generous) financial assistance.

Flexibility could be introduced into grant schemes to permit a balance to be struck between a capital grant and an annual grant, according to circumstances. The overall cost to the public purse need not be affected, but different local targets may require different kinds of incentives at different times. All grant aid could continue to depend on Forestry Commission approval.

The 'mixed economy' of British forestry might be further reflected in an extension of 'shared ownership' schemes. Taking a cue from agriculture, it may well make good sense for one partner to provide the land and another the management expertise. The traditional fixed-term, fixed-rent lease is too inflexible, but other forms of 'partnership' look attractive and there is no fundamental reason why one of the 'partners' should not be the Forestry Commission. On one model, the Commission might have the use of land for one rotation, with annual payments rising as the time for harvesting drew nearer. Other possibilities include agreements under which net proceeds of timber sales could be divided on an agreed basis.

In the last analysis, unless we are to contemplate full land nationalization and bureaucratic management, the achievement of our forestry goals will depend upon land management decisions taken by thousands of private landowners – mostly individuals, but also companies, trustees and other bodies. The only satisfactory way of proceeding is to engage co-operation between public and private sectors in pursuit of national goals. Forestry policy can never be implemented by controls alone: positive management is essential. The need is therefore a clear, positive national policy, which is given incentives and regulations so that individual decisions on the planting, management and felling of trees are in harmony with our national aspirations. That is a tall order, but we could surely get far closer than we are at present!

Ten per cent of Great Britain is now occupied by trees. We turned the corner of forest decline early in this century. Can we not envisage a country towards the end of the next century with around fifteen per cent of its land occupied by trees? That position should not be reached by a narrow-minded concentration on commercial timber production: indeed, it may be argued that concentration on commercial timber production would lead us to even greater reliance on imports! Nor should we rush to plant trees simply because, for the time being, western Europe and much of the developed world have food surpluses. But land *is* now available for trees.

Our forests should grow in response to varied national and local needs, not forgetting the international context. We must remember too that the woodlands and parklands which we prize most highly derive, in many cases, from land use decisions not of this century or even the last one. One of our priorities might be to restore the landscape where we have damaged it – not necessarily by trying to recreate a former scene, but by creating a new one. We are trustees for the future.

If circumstances change and our successors find it desirable not to continue to expand the national forest area, at least we should try to hand on to them a healthy, growing asset characterized by diversity and capable of meeting a wide range of goals. The potential flexibility of forest resources is almost unique and we should capitalize on that.

So the future of our forests and woodlands is one that offers potential for development, as long as we think of innovative ways in which to execute change. It simply does not seem acceptable to develop forestry as a residual of agriculture. We need to be more positive than that. This is particularly so, because forestry, of course, is not the only possible alternative land use in

"Looks as if we're on bloody Wogan's property"

Figure 3.14 Until the Budget of 1988 forestry could be a tax haven for the very rich!
Source: Private Eye

rural areas; other types of development may increasingly take place on our 'surplus' agricultural land. This is made more likely by the freeing up of planning controls in rural areas. It is these other forms, essentially of built development, to which we now turn in chapter 4, in the context of the planning system that determines their extent, type and location.

Chapter 4

A more flexible system for planning and development?

Land-use planning in the countryside

There are fewer reasons than ever for living in urban areas. In the 'accessible countryside', the broad zone up to 100 miles from our major cities, Britain is growing and prospering. For an increasing number, the availability of some form of personal transport has made it possible to combine 'country-living' – with its positive associations of greenness and health, rural simplicity and fine landscapes – with a city job. The commuter from Oxford and Cambridge to the City of London will be matched by his Norwich counterpart by the end of the decade. The English countryside and village life is, more than ever, eulogized in an outpouring of magazines, radio and television series and tourist promotions. The imagery of the friendly vicar, the church spire, and the restored rustic retreat may not be fully reflected in reality, but the lanes and squares are safer and noise, dirt and dust are less likely to invade the scene. With pressures for more dispersed living come demands for higher space standards and better environments at work. In addition, high technology activities and the burgeoning service industries have few intrinsic locational restraints and are more capable of being located in the countryside.

Whilst the tools and procedures of land-use planning in the countryside are perpetually in flux, many of the changes in the 1970s only constituted the 'fine tuning' of a set of priorities that, as we noted early in chapter 1, were established in 1947 and dedicated to preserving agricultural land and the agricultural status quo. With low population growth in the 1980s, a wish by many to leave the cities, and surplus agricultural land, the chance exists, for the first time since the Second World War, to depart from these priorities. Instead of competition for scarce land there is potential space for all in a new arrangement of land uses. With the passing of the old 'consensus' of agricultural fundamentalism, can a new one be formed which will take us successfully to the turn of the century?

In this chapter, we now depart from the primary sector preoccupations of agriculture and forestry that are very much concerned with production levels, economic support and environmental impact. We turn instead to examine the nature of land use or town and country planning policies which are much more closely related to physical controls than economic incentives.

Also, we assess, firstly, the basic tenor of countryside land-use policies inherited from the past and secondly, move on to examine present and likely

Figure 4.1 Image of a village – Selworthy in Somerset. A 'chocolate box' vision cherished by many.
Source: D. Noton

future development pressures on the countryside. Trends in government policy, particularly in respect of new policies for agricultural land, housing and Green Belt are also discussed. We continue with an analysis of the various proposals for reform, made by central government and other interests of importance for countryside land-use planning. We conclude by offering some speculations on the nature of future policy and the possible reform of the land-use planning system itself.

Land-use planning has always emphasized certain policies and interests. One of its main purposes since its inception, has been to control closely the transfer of land out of agriculture into other uses. This existed in the context of minimizing land-take because of the necessity to retain as much agricultural land as possible. Over successive five year periods since 1975 there has been a three-fold decrease in agricultural land loss – from 15,000 hectares a year to around 5,000 hectares as figure 1.2 in chapter 1 shows. At the same time planning controls over agricultural and forestry change – for example, the erection of most barns and silos and tree planting – have not been imposed except to some limited extent in National Parks.

We have seen the operation of policies for the containment of urban development, and the creation of clear-cut boundaries between 'town' and 'country'. These have been pursued through the use of Green Belts and strict development control in the countryside beyond them. By 1984 the area of fully approved Green Belts amounted to 15,800 square kilometres, some 12.5 per cent of England and Wales. The prevention of scattered development, and the restriction of what were regarded as non-agricultural uses of the land, have also been firm policy priorities also assured through development control. As we noted in chapter 1, all of these policies complemented the philosophy that 'every agricultural acre counts'. Conservation groups and preservationists have been reassured by, and have supported, a system which has made extensive use of concepts such as 'amenity', 'landscape quality', and 'conservation'. The definition and continued designation of Areas of Outstanding Natural Beauty (AONB) and Sites of Special Scientific Interest have given additional muscle to the conservationists.

Also in the context of containment, development has proceeded by peripheral accretions to existing towns and rural settlements, as the programme of New and Expanded Towns, prominent in the 1960s, ran down during the 1970s and 1980s. The emphasis on keeping new development off agricultural land has also focused interest on 'infill' in country towns and villages and often considerable increases in housing and industrial densities have occurred. Land-use planners have sought to keep such changes in scale and sympathy with existing development. Government advice to local authorities in the most pressured rural areas of the South East gives a flavour of what is occurring:

> ... the South East can best protect its towns, villages and countryside, if, except for Milton Keynes and South Hampshire, the remaining development takes the form of small additions to existing settlements beyond the Green Belt ... (This) ... would allow a wide range of development opportunities as sought by builders yet enable the needs and constraints of individual settlements to be honoured.

Alongside the maintenance of these consistently restrictive policies for the countryside has been the development of policies for urban regeneration. In the 1980 Local Government, Planning and Land Act, land registers were introduced. These are lists of vacant land held by public authorities which government would like to see disposed of for development. Over 46,500 hectares of such land was listed on registers, one half of which had good

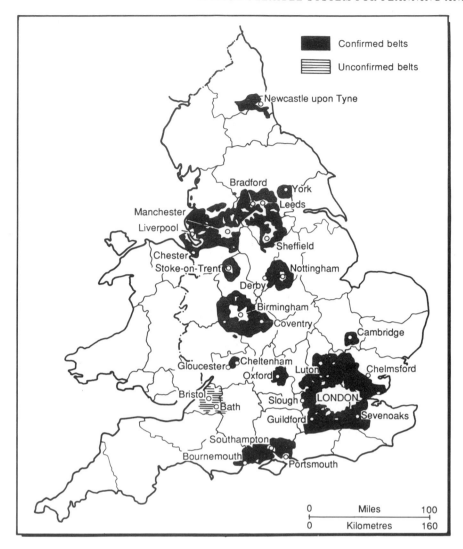

Figure 4.2 England's Green Belts. Designed originally to keep town and country apart, Green Belts are still subject to the most stringent land-use planning controls. *Source*: House of Commons Environment Committee

prospects for housing or industrial and commercial use. They were chiefly in urban areas. Funding for derelict land grants has been increased and focused on urban schemes with a 'hard' development after-use. More recently Urban Development Grants for private sector regeneration projects, and Urban Regeneration Grants have created areas of new development in the inner city. These schemes have certainly helped reduce the amount of unused and under-used land in cities such that we find that only 45 per cent of all new development now takes place on agricultural land.

During the latter part of the 1970s the 'first round' of structure plans – strategic plans at county level – were prepared and approved. All stressed policies of urban regeneration with urban containment. Severe restraints on development in areas of attractive countryside accessible to major conurbations and the coast would assist the urban regeneration process. The local plans, providing detailed development frameworks have, during the past eight years, come to cover over one third of the countryside – with an emphasis on areas of development pressure. Their policies have sought to balance different interests: the desire of many 'locals' (particularly those

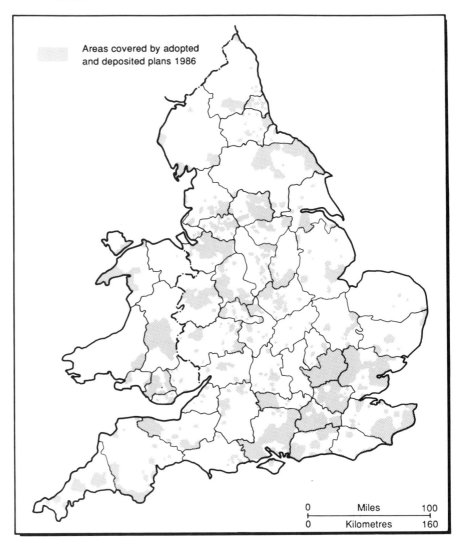

Areas covered by adopted
and deposited plans 1986

| 0 | Miles | 100 |
| 0 | Kilometres | 160 |

*Figure 4.3 Local plans
in England and Wales.*
Drawn up to provide
detailed development
proposals, they are
intended to be consistent
with their county
development plans.
Source: Department of The
Environment

recently arrived in the countryside) for low levels of development; the wishes
of builders for a wide range of sites in locations where they can sell houses,
and demands for greater flexibility in the rate that development can take
place – especially for employment.

The system has, as a result, assumed a combative air and there has been
great reliance on negotiations between planners and developers and a
complex bureaucratic style for making decisions. We have seen the number
of planning appeals rise to over 20,000 per annum by 1987, an increase of
nearly 40 per cent over the year 1979. Many are in pressured countryside
areas, typical cases being the small infill or medium-scale housing schemes in
or adjacent to 'accessible' country towns or villages. Certainly it would
appear that the restrictive land-use policies towards the countryside are
beginning to be challenged.

Groups such as the Housebuilders Federation and other property
developers have, as a result of frustrations caused by delay in getting planning
permissions, criticized the aims of planning as negative and bureaucratic.
Further, academic critics consider planning has been used as a means of
creating spatial segregation. Containment of the cities is, in this scenario,

virtually synonymous with the 'exclusion' of the less privileged from the countryside. Government, mirroring the views of the builders, has also criticized local planning authorities for delay and has instigated a 'league table' of development control performance – published quarterly – intended to improve the speed of handling planning applications. Uncertainties inherent in what, after all, is a system which uses administrative discretion have fitted poorly with an ideology and a government style which 'simplifies' problems. Conservation groups, on the other hand, value the way decisions are made in land-use planning. They contrast the relative openness of our development control process, with its third party rights of comment and participation at appeals, with the deliberations of such groups as Regional Advisory Committees for forestry where major land use decisions can be taken away from the public gaze. We have already suggested in chapter 3 that forestry and woodland decisions might be brought much more closely into the public domain, with a devolution of much decision-making to county councils.

A new geography of location?

Today's planning policies therefore reflect yesterday's patterns of town and country, and the past balance of power between central and local government. Clearly, as the pace of social and economic change is accelerating, (witness the loss of around 2,000,000 manufacturing jobs in the last 10 years), new patterns of land use and settlement are emerging. The leading edge of change appears to be the private sector with its large potential financial leverage. It is therefore important to assess which policies are worthy of retention and how far we can adjust to new market pressures to achieve local community benefits. Should, for example, new economic

Figure 4.4 Executive homes in the countryside. The profitability of the more expensive rural houses in such places as Northleach in the Cotswolds has created a spatial segregation between the more and the less affluent.
Source: ACRE

Figure 4.5 New rural factories. This architect designed building attempts to reproduce elements of the rural image, including church spire and tithe barn. But is it really sympathetic to the rural landscape?
Source: ACRE

activities be allowed into the countryside? Would their possible future growth harm the appearance of the landscape? These issues are closely related to diversifying the rural economy, the subject of chapter 5.

Any package of countryside policies will have to address a range of social and economic changes. Foremost among these we can see a continuing process of *dispersed decentralization*, over long distances, and from the main urban areas. One researcher has identified a *golden belt* or *golden horn* of high population gain across rural Britain. Beginning in Cornwall it runs in a north-easterly direction to Norfolk embracing a range of rural counties such as Devon, Dorset, Somerset, Wiltshire, Oxfordshire and Cambridgeshire (see figure 4.6). In the *golden belt* population growth was 83 per cent of the total for England and Wales over the period 1981–86. Predictions for the 2,001 suggest that in this belt, and the slightly broader area of the *golden horn* (that is, including *all* the shaded areas on figure 4.6) the population gain will be 94 per cent of that predicted for all of England and Wales over the period.

Apart from a general rise in footloose occupations (and some retirement migration), an important generator of this growth has been the creation of jobs in information-based activities and services. Those employed developing such products as electronic components or electronic consumer goods – that is, with 'high-tech jobs' – are frequently found away from the old centres. In the crescent west of London we can see that such growth has its genesis in the pre-war dispersal, in predominantly countryside locations, of military research and development establishments. The development of facilities such as Heathrow Airport, and the M4 and M40 motorways, have only re-inforced this pattern, assisted by planning policies of containment and countryside conservation. A recent study of the high tech phenomenon has suggested that land-use policies have created – from Hampshire to Cambridgeshire – '. . . a limited number of linked, medium-sized growth areas set in a sea of general rural restraint. As a policy of . . . benign perversity, it is just this policy of protecting the character of the countryside

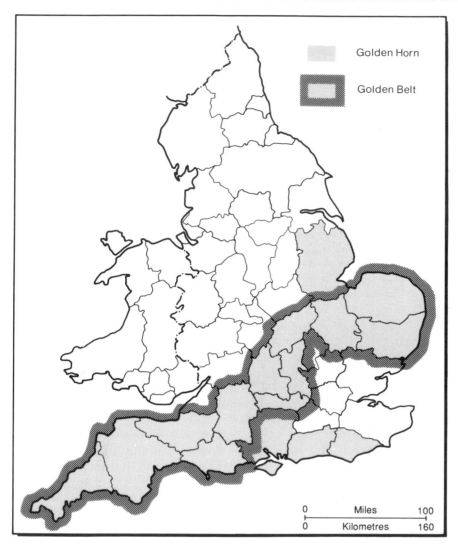

Figure 4.6 Golden Horn and Golden Belt. Two spatially defined areas subject to considerable population growth and development pressure. *Source:* P. Hall

which has made the area so environmentally attractive to both indigenous and incoming high tech firms.'

As some of the main wealth generators in the economy decentralize, important spin-off effects arise. Service jobs for the growing outer city populations create pressures for office development in medium-sized country towns. Attractive town centres may therefore be under threat. Major shopping chains also seek to move 'out of town' to 'capture' the migrating expenditure. By 1987 over 60 million square metres of large out-of-town retail schemes were at the proposals stage, including at least a dozen major superstores around the M25. There were also half that number around Greater Manchester and Exeter in the South West. One giant proposal at Runnymede in Berkshire would, if constructed, include 93,000 square metres of retail space in four decks located in a country park setting, (with windsurfing, boating and fishing activites), together with space for over 8,000 cars. A proposal at Blue Water Park, Dartford, if approved, would have a cable car bringing shoppers to the centre of a superstore constructed on a quarry floor about 40 metres below the general level of the Kent countryside.

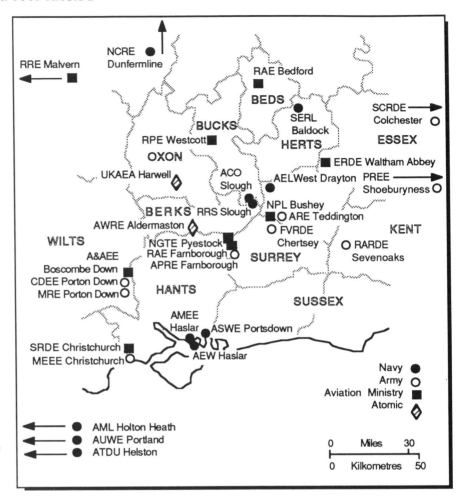

Figure 4.7 Military research and development establishments. Some would argue that the location of government activities in the South East has acted as a further spur to economic growth.
Source: Select Committee on Science and Technology

Shopping in such locations may well become a new form of Sunday outing for countryside leisure.

Infrastructure investments are also leading to new countryside pressures. The Channel Tunnel Act and the creation of new rail facilities, including a major interchange at Ashford, have created a new impetus to development. Kent County Council are, as a result, proposing major relaxations of restraint policy up to 60 miles from the Tunnel entrance. Similarly, in the West Midlands, the M42 axis and Birmingham Airport are focii for growth. In South Manchester the Airport is also regarded as a regional-scale catalyst for new economic activity.

The urban fringe is under particular pressure. The Countryside Policy Review Panel in its Report *New Opportunities for the Countryside* see a widening belt of land of uncertain character between the countryside and the town. Hertfordshire, where agricultural land is being lost at the rate of 2,000 hectares a decade, is an example cited;

> ... In the southern part of the County extensive stretches have half of their rural area in holdings not now classified as 'agricultural' by the Ministry of Agriculture. They include recreation sites, 'hobby farms', public utilities, airfields, residential and research institutions, scout camps, reservoirs and mineral workings, as well as commons, wetlands

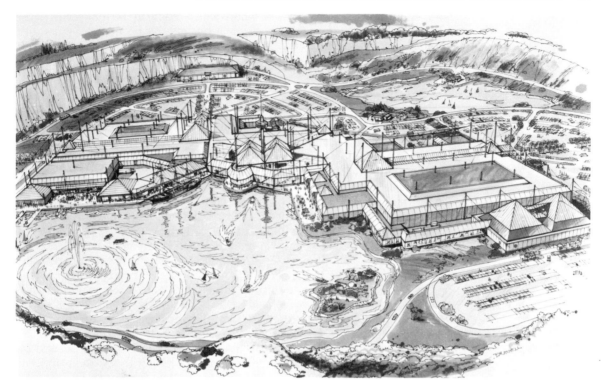

Figure 4.8 A new concept in shopping and leisure facilities. The Blue Water Park proposal at Dartford in Kent integrates both types of activity. This is becoming an increasingly common theme in large-scale planning applications. *Source*: Shearwater Property Holdings PLC

and areas set aside as wildlife reserves. *Most of it looks like countryside but it is not agricultural in character*. (Editors' emphasis.)

There are clearly extensive areas of despoiled and degraded urban fringe countryside and there is a pressing need for a new policy to sustain the countryside environment near towns.

Some of the land-use side effects of the need to control agricultural surpluses have created a further axis of concern. As agricultural incomes come under pressure the need to diversify land use has become more insistent. Thus farmers who 10 years ago sought protection from the planning system, in that it insulated them from urban influences, are now keen to promote a variety of new business activities. Many of the commercial ventures – whether manufacturing items such as craft goods, ice cream or yoghurt; sport, recreation or other forms of retailing – will require planning permission. Success in this regard has varied. Certainly, some local residents living near to such proposals, have resisted development which makes the countryside noisier and busier, particularly if they are recent arrivals in the area.

A recent report *The Challenge of Change* by a study group of the Royal Town Planning Institute suggests that our present planning concepts, with the emphasis they place on the *physical* control of development, fall far short of what is needed to cope with development trends. The Countryside Policy Review Panel warns that unless the demands to move to a more dispersed pattern of urban growth are carefully managed, they could have an unwelcome impact on the countryside, just at a time when the rural and small town environment is being increasingly valued. Planners see agricultural land values rising in the hope of new urban-style use being given approval. The idea of 'Spread City', relying on high mobility, dispersed patterns of work and limited forms of self-help may be where we are going in the long term (figure 8.13). But our route to such a destination may by no means be smooth. Policies now being implemented, and to which we now turn, will largely determine the pattern of town and country to the turn of the century.

The political response to locational change

Successive Conservative administrations have, since 1979, followed a range of ideas which appear contrary to the retention of strong countryside planning based on regulatory powers. Economic philosophies put forward have involved ideas of 'rolling back the boundaries of the State' to free individuals to act as entrepreneurs thus increasing production and creating jobs. This individualistic flavour has been re-enforced by policies to increase levels of private home ownership. Nearly one million council houses have been sold and support for public-sector housing provision has been cut by 50 per cent. The aim has been to stimulate the economy by 'tax cuts', and the privatization of major industries (for example, British Telecom and the water authorities). A wider range of economic agents are to be exposed to market processes. Strict monetary controls on the public sector reflect a mistrust of local authorities, (and of officials who exercise 'discretion' in a rather nebulously-defined public interest), as well as the search for value for money in service delivery.

As land-use planning powers in the countryside are centred around the concept of administrative discretion it is not difficult to see their unpopularity. Although one government ginger group has suggested discretionary development control system should be replaced by a more law-bound broad zoning system for environmentally sensitive areas, with only the 'laws of nuisance' elsewhere to protect the impact of one user upon another, a less radical pattern of change has emerged. There have been moves to reduce remaining locational constraints on land uses, and to support specific groups such as private housebuilders, small industrialists, and large retailers. At the same time attempts have been made to re-assure existing rural residents seeking retention of the *status quo* that 'essential' regulatory tools are not being jettisoned.

The major policy adjustments have been made by re-writing advice in Circulars (we have introduced the more important of these in chapter 1) and by altering submitted policies in development plans. This latter process has been a regular feature of planning since its inception – local authorities have always proposed more restrictive countryside policies than central government have been prepared to approve.

Economic activity in a new climate

The emphasis in terms of industrial and commercial development has been on 'lifting the burden' with respect to location and choice of sites. The government has also attempted to persuade local authorities that a wider range of economic activities, over and above agriculture, are suited to rural locations. The 1980 Circular *Development Control-Policy and Practice* (22/80) suggested that disused farm buildings could be re-used for employment purposes and could conform with the rural scene. At the same time, it was argued, the introduction of new jobs would 'prevent loss of services and keep a viable and balanced community'. The 1984 Circular *Industrial Development* (16/84) stressed that many commercial activities could be carried on in rural areas without causing unacceptable disturbance. Local authorities

Figure 4.9 New employment in the countryside. Recent government circulars have done much to liberalize planning policies and encourage new job initiatives in rural areas.
Source: Rural Development Commission

concerned about intrusion were reminded that many light industrial uses are less noisy than some activities related to agriculture.

Evidently concerned that development plans covering the countryside had been too restrictive in the past, local authorities have been encouraged to develop more 'flexible' policies. The 1984 Circular *Industrial Development* (16/84) suggests that plans 'can rapidly become out of date' thus implying that the approval of schemes not in such plans can be more easily justified. Local authorities, it suggests, should not turn down employment proposals in an attempt to steer them to land allocated in these plans. There should be a wide choice of sites of different sizes and with different facilities in each local authority area for businessmen to choose from. Firms using the new technologies should, particularly, have priority and sympathetic treatment.

Basically the advice reiterated in the 1986 Circular *Development by Small Businesses* (2/86) has reflected what is already going on around the country. In the 1980s local authorities have relaxed policies to allow more employment uses into the countryside. Emphasis has been placed on the re-use of existing buildings; for example redundant farm buildings or large country houses which would be uneconomic in single residential use. In addition, large numbers of 'starter houses' have been provided in villages in rural pressure areas. Other 'loopholes', such as the development, for employment, of airfields with existing user rights have also been exploited. Research commissioned by the Department of the Environment on rural employment suggested that the planning system 'had not represented a severe restraint on the development of small businesses'. The main types of proposal refused were for activities such as car repairing, car breaking, the storage of scrap metal and transport activities (such as haulage contracting). These are often considered to be major intrusions into the rural scene.

Freeing restraints on housing

Private housebuilders argue that they require a continuous flow of new sites in marketable locations. The sites with fewest physical development problems are the accessible countryside near to towns. In a region such as the South East decreasing household size, (manifest in the need for single parent family and old person's accommodation), leads to estimates of a requirement for around 750,000 plots to the turn of the century. Much of the demand will fall in to Green Belt areas, AONBs or similarly restrained rural locations. Shire county authorities are reluctant to breach existing policies to allow for large numbers of new houses to satisfy such predictions.

The conflict between local ideas of what is attractive countryside and how it should be protected and government policies to increase private home ownership, is therefore clear. In order to speed the release of new land, policies in Circulars have been changed. In 1980 the housebuilding industry had been allowed a 'place at the table' conducting joint land availability studies with local authorities. Builders could therefore assess whether proposed housing sites were 'marketable' or not. If considered unmarketable they could be replaced by more attractive sites for inclusion in development plans.

A 1984 Circular *Land for Housing* (15/84) further adjusted the definition of land 'availability'. Previously it had been sufficient for each district to have a five-year supply in a plan based on local authority–housebuilder joint studies. The new five-year supply should now 'cater effectively' for the demand for housing in localities. Each district planning authority would have to justify that at least two years' supply was immediately available (that is, land on which development could start tomorrow). Allocated land had also to have a 'reasonable prospect of willing sale' to be regarded as available. All

Figure 4.10 'We've yet to hear Prince Charles' comments on this particular development.'
Source: Chartered Surveyors Weekly

this pointed to keeping open a continuous flow of land for development irrespective of rural locality.

As far as location is concerned the *Land for Housing* Circular suggested that 'expansion of a town into the surrounding countryside is objectionable on planning grounds if it creates ribbons or isolated pockets of development, reverses accepted policies for separating villages from towns', or contravenes Green Belt and AONB provisions. Nevertheless the Circular suggested that most new housing would continue to be on new sites and that development plans would need to build-in flexibility in the rate at which land is made available for housing. Phasing was seen as 'arbitrary' and, apparently, forbidden. In the first Draft of the housing Circular it was suggested that in some areas it may exceptionally prove the best solution to plan for 'new settlements' rather than to expand existing communities. Following protests from the Royal Town Planning Institute and some of the local authority associations, among others, the final advice suggested that proposals for 'new settlements' would normally go through the statutory plan making process. This would result in their evaluation against other methods of fulfilling housing requirements, for example, by the rounding off or infill of existing settlements.

Adjusting green belts policies

Attempts have also been made to encourage local authorities to set Green Belts well back from urban areas so that the need for future incursions can be kept to a minimum. The Draft Circular on Green Belts, issued in 1983, caused consternation among local authorities, conservationists and rural community groups. First of all, it stressed drawing back the boundaries of Green Belts defined in new development plans to the long-term limits to development. (Although governments have always been reluctant to put a time limit to this, it was clear that long term would mean well beyond the period of the development plan). Secondly, it suggested that any land between the 'five year' development boundary and the 'long-term' Green Belt should (in the absence of phasing) be protected by 'the normal processes of development control'.

The government also suggested that authorities should consider omitting Green Belt notation from 'relatively small detached areas of land among existing development', incorporating other land at the outer edge of the Green Belt to compensate for the adjustment. Some saw this as a new form of 'rolling green belt' – no less than the introduction of a system of flexible urban development boundaries.

This prescription, combined with that for housing, appeared to give even greater precedence to housebuilders' market assessments, and to invite local authorities to 'roll back' the Green Belt. There is no doubt that pressure from back bench MPs, complaints from conservation groups such as the Council for the Protection of Rural England and the holding of an Inquiry by the House of Commons Environment Select Committee, moderated Ministers' intentions for the Green Belt. As a result the approved Circular relating to Green Belts (Circular 14/84) stresses links between Green Belt restraint and urban regeneration, and suggests approved Green Belts will only be altered in exceptional circumstances.

These new guidelines have done little to remove the obvious conflict in many rural localities between allocating sufficient land for market demand whilst retaining pre-existing Green Belts. At a succession of appeals and hearings into structure plan alterations the conflict has been played out. Counties tend to place a low figure for the number of housing plots required for the

1991–2001 period, based on satisfying local demands. This is countered by far higher figures, using more generous assumptions, produced by the house builders. At the hearings into the Hertfordshire and Surrey alterations in 1986 and 1987 it has been argued by the local authorities that there is no more capacity in many towns and villages without trespassing onto Green Belt countryside. In both counties it has been suggested that surplus hospital land, currently in the Green Belt, could exceptionally be used to help make up the difference. However, again under pressure from conservationists and shire Tory MPs, a specific 1987 Circular *Redundant Hospital Sites in Green Belts* (12/87) was produced to clarify the government's position. This stated that if the existing buildings could not be used for institutional use they could be used for 'other suitable uses'. If redevelopment rather than conversion was the only possibility, the redevelopment should not occupy an area larger than the original hospital buildings. Proposals for the Shenley site in Hertfordshire have been supported by Hertsmere District Council following a feasibility study by economic consultants, but they appear to contravene the guidelines in the Circular in that they propose development of more than the original hospital area. The DoE are suggesting that this proposal is not sufficiently 'exceptional' for the Green Belt to be breached. A proposal by Surrey and Epsom and Ewell Councils that 2,000 houses should be constructed on Green Belt hospital land near Epsom, however, is likely to receive government approval within the Surrey Structure Plan.

New settlements in the English tradition?

It would seem then, that the new geography of location coupled with a political philosophy that leans much more towards the free market has led to a deal of liberalization of land-use planning policies. This has been particularly so in the case of industrial developments, housing and Green Belts. But on a slightly larger scale, pressures have grown for the development of whole new settlements.

Concerned by what they saw as a log-jam of indecision, the major private housebuilders in 1983 set up a development company to build major new 'country towns'. They thus sought to influence the government to change the planning system in their favour by challenging settlement policies head-on. Consortium Developments Ltd. are proposing a number of freestanding country towns which they describe as 'communities in an English tradition', a new mechanism to meet housing demand at the same time offering an alternative rural lifestyle. New country towns would offer the inhabitant and the visitor 'a sense of positive balance between city and village'. Each town would have about 5,000 homes; a population of between 12,000 and 15,000 and 340 to 400 hectares would be allocated for housing, employment and commercial use, health, education, recreation and public open space.

Creating 15 to 20 of these settlements would, the builders contend, help reduce pressures for infill, urban sprawl and piecemeal development in and around existing towns and villages, as well as reaping 'benefits of scale' in construction terms. Figure 4.11 shows the locations of the first Consortium Developments' proposals and a number of similar schemes in the South East. Each scheme, which may include some plots for small builders and allow for various tenure arrangements (for example self-build and housing associations), has been varied slightly to fit the local topography. The imagery being marketed is that of rural life in a village of vernacular design. One prospectus shows bullrushes fronting, and ducks overflying, a view of the town centre. Where houses can be seen they appear generously arranged in surroundings reminiscent of the garden cities of the early years of the century (Figure 4.12).

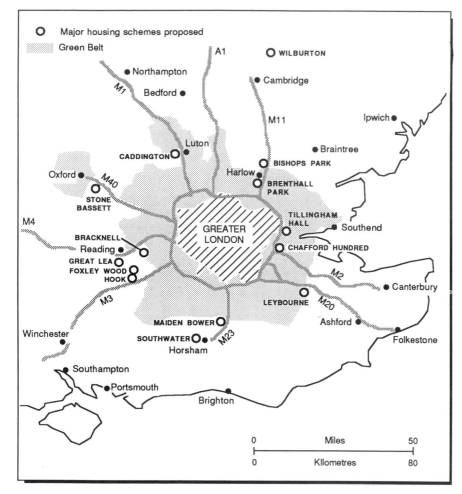

Figure 4.11 New country towns in the South East. Planning applications for whole new settlements in the countryside are on the increase in the wake of a relaxation in planning controls.
Source: J. Blunden and N. Curry

The first scheme of Tillingham Hall, in the Essex Green Belt near Basildon, was refused on appeal by the DoE. The Inspector suggested that although evidence of a shortfall of new provision for housing could possibly be substantiated for the latter years of the century, there was no reason to believe that the local authorities could not release sufficient land in other locations. The housing situation was not seen as exceptional enough to overturn the general presumption against development in the Green Belt. However, the refusal letter left possibilities open stating that '... well conceived schemes of this kind, in appropriate locations, may have a part to play in meeting the demand for new housing.'

As a result schemes for Foxley Wood, beyond the Metropolitan Green Belt in North East Hampshire, and Stone Bassett, adjacent to the Oxford Green Belt, are being most actively pursued. The 'fashion' for large mixed developments of this sort has reached such a level that twelve proposals for new country towns are being considered around Cambridge, and there are a range of proposals along the M4, and the main Channel Tunnel transport corridor through Kent.

New attitudes to agricultural land

In chapters 1 and 2 we have outlined and evaluated a number of measures currently under discussion to counter problems of food production surpluses. One of the main planks of the countryside land-use planning system has been the 'farming first' or 'every agricultural acre counts' presumption. Any change of this longstanding priority, which involves consultation between local authorities and the Ministry of Agriculture, Fisheries and Food (MAFF) over the quality of farmland (essentially measured by a five point agricultural land classification), could be expected to raise lively controversy among the development and conservation interests involved.

This controversy certainly came with the setting up of ALURE, and the production of the Draft Circular, *Development Involving Agricultural Land* . As we have noted in the Preface to this volume, there were clear tensions between MAFF and the DoE before even government agreement was reached about the extent of 'freeing up' the countryside. The Draft Circular *Development Involving Agricultural Land* suggested that the long-held presumption against taking agricultural land of grades 1, 2 and 3 should, in future, cover grades 1 and 2 only. The threshold for referral of proposals to MAFF would be 20 hectares and not the four hectares which had previously applied.

Outside Green Belts, National Parks and AONBs the 'farming first' presumption in development control would be replaced, it suggested, by an assessment which considered *as of equal importance* 'the agricultural implications, together with the environmental and economic aspects' of a development. The need to 'promote economic activity that provides jobs' would be a material consideration, as would the need to control the rate at which land is taken for development. Outside the main countryside designations local authorities were advised that other areas of 'good countryside should be conserved and protected from development'. A more lenient attitude to the removal of occupancy conditions on agricultural dwellings was also proposed; 'realistic assessments' should be made of the need for them. This opened up the possibility of new forms of gentrification parodied in figure 5.4.

This prescription was interpreted by Labour's environmental spokesman, David Clark, as a 'charter for speculators'. The Council for the Protection of Rural England in a press release 'new storms ahead for government in farmland row', suggested that reducing the protection given to most farmland, without providing a strong compensating rationale for countryside protection, would trigger damaging erosions of the countryside by developers 'in a way that Ministers currently claim not to want'. They mistrusted a Ministry who had upheld only 29 per cent of 9,000 appeals determined in 1979 but, by 1985, was upholding 40 per cent of the far larger number of 14,000.

Figure 4.12 Designing a new community. Foxley Wood in Hampshire is one of the more imaginative of these new schemes. This artist's impression shows the central area of the proposed development. *Source*: Consortium Developments

Certainly the proposals appeared confused. First, it appeared that as only 13 per cent of the farmland of England and Wales is in grades 1 and 2, and given the high threshold proposed, MAFF involvement with the local planning process would be virtually eliminated. Around 10,000 hectares a year went through this consultation process over the period 1977–80. Second, it was unclear what constituted the 'good countryside' that should be conserved. Who would decide its status? Furthermore, how would the need to control the rate of land release fit with exhortations in other Circulars, such as that in *Land for Housing* in 1984, to allow wider flexibility in the rate of land release? Also, in what circumstances would the need to promote

LAND FIT FOR TORY OWNER OCCUPIERS TO DWELL UPON

Figure 4.13 A charter for speculators? David Clark's view of government proposals to relax planning controls in the countryside. *Source:*

economic activity override the need to retain good countryside? Little in the way of elaboration was forthcoming.

The approved Circular *Development Involving Agricultural Land* (16/87) contained minor changes, the content of which had been checked informally by Ministers with the conservation organizations. These included the replacement of the statement referring to 'good countryside' with a criterion of 'the continuing need to protect the countryside *for its own sake* rather than primarily for the productive value of the land' (Editors' emphasis); the extension of the consultation arrangements to grade 3a land, although the threshold was retained at 20 hectares; additional stress on the need to re-use vacant and derelict land in urban areas; and an emphasis on having regard for the 'provisions of the development plan' in making decisions on individual projects.

The new Circular now asserts that the best and most versatile land (presumably grades 1 and 2) has a special importance as a national resource for the longer term. Given this apparent re-statement of the importance of good quality agricultural land, and the need to protect other land (presumably of lower quality) 'for its own sake' perhaps we have come full circle.

Writing in the *Housebuilder* magazine one agricultural consultant suggests '... overall it is hard not to conclude that the new Circular leaves available to local planning authorities more rather than fewer reasons for refusing applications on green field sites'. If this is the case it will be the opposite of the outcome intended by the government and predicted by conservationists. Certainly the clarity of the original intention – to give less emphasis to agriculture in development decisions – appears to have been obscured.

Proposals for reform

There is undoubtedly widespread disquiet with the way our land-use planning system is currently operating in the countryside. Overburdened with roles (providing the infrastructure for development land, conserving valued environments, co-ordinating development, mediating between outside and local interests), its outputs satisfy few. There is widespread general agreement, however, over the need to retain some form of public intervention to counterbalance the worst impacts of market processes. Beyond this progress proves difficult. During 1986 at least seven bodies; the British Property Federation, the Royal Institution of Chartered Surveyors, the Nuffield Foundation, the Confederation of British Industry, the County Planning Officers Society, a Study Group of the Royal Town Planning Institute, and the DoE all put forward proposals for change to the existing system of land-use planning.

Their prescriptions covered both the policies, and the instruments of planning. Development-oriented groups such as the Royal Institution of Chartered Surveyors criticized planning for its 'lack of clear, defensible and robust' policies and, what it terms 'the excessive political influence at local level' on planning decisions. The British Property Federation suggested a simple zoning system for the whole of Britain which would effectively remove discretion from local hands. The Royal Town Planning Institute study *The Challenge of Change* suggested that land-based policies should be altered radically to enable the further dispersal of the population on the basis that there is room for all without injuring the agricultural industry. At the same time they saw the need for the benefits of dispersed living to be extended to the less prosperous sections of the population.

The Report by the Nuffield Foundation is perhaps the most far reaching and exhaustive. It catalogued a deep-seated failure of powers, purpose and results, stating 'planning has commanded neither the powers nor the authority to influence decisively the pattern of economic change, or the distribution of wealth'. Its recommendations include a far higher profile for public intervention in countryside planning. Planning actions should extend from its narrow land-use base to embrace social and economic considerations. The role of local authorities as 'guardians of the local environment' should be extended. Development control should cover farm buildings, building extensions and farm roads. In local authority-designated 'landscape control and management areas' alterations such as hedge removal or major drainage works on agricultural land would require planning approval. Discrimination in favour of local occupiers of housing and other developments should be allowed where no national interest is affected.

The above policies would operate within a system where triennially-reviewed development plans would be the norm, operating under national planning guidelines (published by the DoE), informed and up-dated by an annual White Paper on planning and the environment. Stress is also placed on the need for consistent statements of *national* policy, but for central government to also allow more policy variety at local level. Intervention locally should only occur where a matter is clearly of national importance.

The DoE's proposals on the future of the planning system are narrower in scope, and are largely related to alterations to procedure. They do, however, represent potentially the most important set of changes to the structure of planning since 1968. Claiming to be setting out to 'retain and strengthen the basic elements of the system' and to improve its efficiency and effectiveness, a simplistic critique of the present system of development plans is outlined.

Structure plans are seen as too lengthy, too detailed and having taken too long to complete. The relationship between structure and local plans is seen as confused with frequent overlap. The procedures for local plans, with their two rounds of public participation, are also seen as lengthy and complex. Non-statutory policies and plans (many of which relate to the countryside and natural environment) are, it is asserted, also adding confusion to the decision-making process. The analysis is very much a 'view from the centre' and many of the proposals would aid 'manageability' (or central control) by civil servants and Ministers, reduce public participation, and retain the narrow focus on land use criticized by many. Looking closely at the recommendations it is difficult to see other than an emasculated role for local authorities. Under these proposals regional guidance is to be advisory and produced by the DoE. The arrangements, which are now in place in most parts of the country, involve the local authorities preparing technical reports and policy recommendations on topics defined by the DoE. (In the West Midlands, for example, 'countryside' is not a topic but Green Belt, land for housing and urban regeneration are!). At a major meeting or 'conference', local authorities hammer out their differences and put the proposed advice to the DoE. Only at this point is there wide consultation. The final content of the advice is also at the behest of the Ministry. The advice need not cover all regions of the UK or the whole of any of the Standard Economic Regions. It is up to local authorities to come together and convince the DoE that the exercise is of value for their particular area. Evidence emerging from the advice under preparation suggests it will be extremely general, and will give little guidance on the future of the countryside.

The proposed 'County Statements of Policy' would also be advisory being prepared by the counties on topics defined by the DoE. The topics prepared would include land for housing, transport and strategic highway matters, as well as protection of the countryside. The draft county statements would be

1. Regional guidance to be issued by Government following consultation with local authorities.

2. Structure plans abolished.

3. County Statements of Policy to be introduced.

4. District Development Plans to be sole statutory development plan for each District.

5. County-prepared minerals plans to form part of District Development Plan.

6. Rural Conservation Areas to be introduced;
    general extent to be determined by Counties

Figure 4.14 The future of Development Plans. These are some of the main proposals of the Department of the Environment.
Source: Department of the Environment

'examined' at hearings conducted by the county. The policies proposed by the county could then be amended by the Minister before finally being put in place.

The only formal development plans would be District Development Plans. Normally prepared by districts, there would be one for each local authority in England and Wales. They would contain all urban and rural land use policies in operation in the local area. The local planning authorities would have to decide whether the non-statutory development control guidance now in use (for example, on aspects of conservation, landscape improvement or countryside management) should be put in the plans or dropped altogether.

In the DoE proposals Rural Conservation Areas would be defined areas of the countryside where the environmental quality or natural or historic character is of sufficient importance locally to justify different development control policies. The general extent of a Rural Conservation Area would be defined in County Statements with the definition of detailed boundaries left to the District Development Plan stage. The DoE's idea is to replace the plethora of local definitions such as Area of Great Landscape Value, Landscape Conservation Area, or Area of Semi-Natural Importance with the Rural Conservation Area. Creating a consistent nomenclature in this way would, it is claimed, 'provide a clear and coherent framework within which rural conservation policies can be positively promoted and which the public and developers can readily recognize'. No extra powers would exist in such areas; there would merely be a different set of presumptions regarding the criteria for development control. However, confusion appears to have been added by the view in the same document that Rural Conservation Areas could cover a wide variety of landscapes (thus apparently begging the question of whether they also replace areas of importance for nature conservation that are not otherwise statutory defined); and that the policies across Rural Conservation Areas could differ thus making them a less easily recognizable special area for the public and developers than hoped.

Rural Conservation Areas have received little support. The Royal Town Planning Institute, the Council for the Protection of Rural England, the Countryside Commission and the Nature Conservancy Council have all rejected the idea at the consultation stage. Most of the countryside interests responding to the proposals considered the definition too weak, rendering the effect of such a measure trifling. Others thought it would be confused with other 'designations' such as AONB, or would downgrade the attention given to conservation in other parts of the countryside.

By early 1988 the regional guidance for the West Midlands is virtually finalized, and the process in four or five other regions is well underway. No legislation was required for its introduction. Rural Conservation Areas are being 're-thought' within the Ministry. The other measures remain firm intentions, but their introduction has been deferred until the 1988–89 Parliamentary session.

Perhaps not unconnected to a groundswell of opinion that national policies should be clarified, the government, in January 1988, published the first of a series of *Planning Policy Guidance Notes*. These provide general policy guidance on matters such as land for housing; industry, commerce and small firms; rural enterprise and development and simplified planning zones. The Circulars from which they have been distilled remain in force for the time being. The idea is that advice on legislation and procedures will increasingly be found in Circulars, with *Planning Policy Guidance Notes* being the main source of policy guidance (figure 4.15). If these Notes on the present model, were to become the sole guidance for countryside policy, they

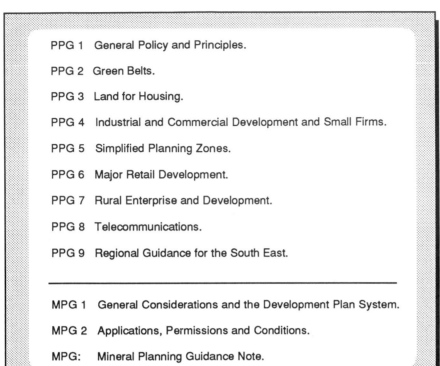

Figure 4.15 Planning Policy Guidance Notes.
Source: Department of the Environment

would constitute a far more generalized set of advice than even that found in current Circulars.

Towards 2001

New circumstances have disturbed the old consensus over countryside priorities. In an effort to loosen controls on economic activity generally there have been exhortations to planning authorities in rural areas to keep up the flow of new sites for housing and employment. These have had some success. New settlements, perhaps on the Consortium Developments model, are however strong contenders as a complement to the existing policy of urban peripheral development. Large mixed developments – combining out-of-town shopping, leisure and office development – may increasingly invade the accessible countryside. Certainly new 'looser' settlement forms are around the corner. Such developments will do little to provide affordable housing for locals working in rural industry; a situation which has been made worse by the mass sale of rural council dwellings since 1980.

We can expect, in the de-regulation scenario, a new multi-activity country-side. What was an agricultural landscape will be transformed by new enterprise. Small industry, leisure activity, forestry and service activity will provide a new veneer to the visual scene. Major questions remain, however, over how far farmers themselves are capable of diversifying, and whether some of the forms of diversification will be acceptable to locals who do not depend on agriculture for their livelihoods. We will return to this question in chapter 5.

It is possible we may see new forms of rural gentrification. The land 'set-aside' from agriculture may be bought by intermediaries and sold in 2 or 4 hectare lots. New forms of low density urbanization – already evident in the countryside around Belfast due to policy relaxations in 1982, and around Dublin in Eire – may be ahead. Thus, not only may we see a more mixed activity countryside, but we may also see an increasingly non-agricultural one.

The scale of countryside change seems to suggest a need for a strengthening, not a weakening, of imaginative and sensitive regional planning. Priorities need to be sorted out at a level below that of Whitehall, but above that of the average district. For this reason the proposals of the Countryside Policy Review Panel for rural development strategies, drawn up at county level and having the same scope broadly as National Park management plans could be important. There has been progress in this respect in some counties (for example, Hampshire) but success will involve setting up machinery to mesh more closely the decision-making activities of landowners, local communities and the government agencies. The introduction of a ministry coordinated 'urban fringe policy' to steer agricultural and environmental changes in the most vulnerable areas, would also be a basic tool for progress. However, present government changes are likely to 'hollow out' the central decision-making ground by removing many of the residual policy-making powers of the counties contained in structure plans. There is also a trend towards channelling money through area-based schemes administered by quasi-government bodies, sometimes with low levels of public accountability.

The need to establish clear priorities in land-use planning involves developing a logical hierarchy of decision-making, as well as policies to ensure effective conservation, the encouragement of agriculture and the provision of jobs. In a world where deregulation and individualism are the watchwords it remains unclear how successful local authorities will be in managing change through a mixture of rural development plans and the pre-existing system of development control.

So it looks as though Professor Dennison's vision that we mentioned early on in chapter 1 as being a minority, almost heretical view in 1942 – that of a countryside of diverse economic interests populated by those other than the agricultural population – may at last be coming about, at least in lowland Britain. But will this 'new development in the countryside' and the intended reforms to the land-use planning process simply benefit the already better off? This is an issue we return to again in the final chapter of this volume. But much will depend on the way in which we diversify the rural economy given that the land-use planning system is set to tolerate a broader range of developments. We examine this process of diversification in chapter 5.

Figure 4.16 'So where do we go after the Green Belt?'
Source: Planning

Chapter 5

Diversifying the rural economy: the real stimulus to rural wealth?

Opportunities for diversification

The major impetus for rural diversification is based on the coincidence of two policy changes. The need to curb agricultural production and reduce the amount spent on agricultural support which we have discussed in chapter 2, and the 'liberalizing' of the land-use planning system which we have assessed in chapter 4, combine to create opportunities for diversifying the rural economy which are without historical precedent. Although it is not yet clear how these two sets of forces will interact to produce a rural economy in the 1990s and beyond, what is certain is that major changes will occur. In exploring what these might be, in this chapter we examine some of the underlying themes that are promoting rural diversification. It is here that we look at a number of underlying principles driving the policy changes that we have discussed so far in this book.

The first of these is the changing values of, and attitudes towards the countryside. Unless we understand the collective psyche of all the relevant parties in the process of economic change, the observable facts of change have little meaning. The farmer's decision to diversify, the redundant entrepreneur's desire to establish a small business, or the planning committee's decision to reject a development proposal can be put into context if prevailing attitudes are understood. There are two main areas in which attitudes are changing or are likely to change: firstly, attitudes to what constitutes appropriate economic activity on farms and in the countryside; and secondly, attitudes to work and employment.

A second theme we explore in this chapter in the context of diversifying the rural economy is to examine the facts of change and observe the adjustments that have occurred and are occurring in the rural economy. This will help to explain the declining role of agriculture and other primary industries and the expansion of manufacturing and particularly service employment in rural areas.

What has happened has been influenced to an extent by government policy. We have examined agricultural policy, forestry policy, regional policy and land-use planning policy in chapters 2, 3 and 4 and these have all affected the process of employment change. We must now as a third theme of this chapter explore the consequences of particular policies, specifically for employment change, and endeavour to understand the absence of any unifying influences until very recently.

In the late 1980s new initiatives are being pursued, some of which have been collaborative efforts by various public agencies. We examine these as a fourth theme in this chapter and ask a number of questions about their potential. Can a unifying gel be found? Will these new initiatives succeed? Are some of the elements of the ALURE package merely lip service to wider economic and environmental concerns, with grants too inadequate to coax the most progressive farmer into action? We must review what is happening and try to identify the ingredients of a vibrant rural economy. But must we look forward to the 1990s haunted by images of the depressed countryside of the 1930s, or can we foresee a different and hopefully better future for our countryside? It is this perspective that provides the final theme in this chapter.

Changing values and changing attitudes

Figure 5.1 Agricultural depression in inter-war Britain. Such scenes encouraged the Scott Committee to seek a more prosperous alternative for the post-war countryside.
Source: MERL

Different people with an interest in the countryside do not necessarily hold identical views about what are appropriate economic activities to be pursued in the process of rural diversification. For a variety of reasons those who have influenced countryside policy in the previous forty years have tended to want to see rural areas through green-tinted spectacles. Initially the tint was an agrarian green; later it became an ecological green. As we have seen in chapters 2, 3 and 4, the attitudes and values that have underpinned these green-tinted visions are currently being subjected to a sustained challenge. The outcome of this challenge from those whose ideas are rooted in economic individualism will strongly influence future employment opportunities in rural areas, a point we return to later in this chapter.

As we noted in chapter 1, the dominant ideas in shaping the economic

activity and the employment prospects in the post-war countryside can be found in the 1942 *Report of the Committee of Land Utilisation in Rural Areas* (the Scott Committee). The majority report of the committee noted the decline which had been endemic in rural areas in the 1920s and 1930s and tried to outline the elements of a vision for the future. We should not forget that the wartime conditions in which this vision was constructed were far from normal.

The principal ingredients in the post-war strategy were based on positive support for farming and traditional rural industries and on negative controls on alien land-use developments that would 'threaten' rural life. Both elements were entirely understandable. The dereliction and decay in British farming between the wars had created an industry ill-prepared for the heavy demands that war brought. Subsequently, protection for farmers from the turbulence of the free market seemed like a reasonable insurance premium, a fair price to pay for a guaranteed supply of food. The control of alien development was also understandable in the light of the largely uncontrolled development in the inter-war years which included seaside shanty towns, squatter settlements, suburban sprawl and new industries.

But Scott's vision was based on retrospect, not prospect. And as we noted in chapter 1 and at the end of chapter 4, the dissenting voice of Professor Dennison on that committee showed that the flaws in Scott's vision did not need hindsight to make them apparent. Scott's vision of a countryside, pivoting around a prosperous agricultural industry, that was both attractive and affluent, full of well-fed, well-housed workers and rustic craftsmen, was simply unattainable. To become efficient, agricultural labour would need to be replaced by machines and rural craftsmen by urban factory workers. Agriculture as an industry needed to undergo a new agricultural revolution which would destroy the cosy intimacy of the Scott vision. Dennison's might have been a lone voice on the Scott Committee, but a study of rural planning needs in Oxfordshire in the late 1940s that was carried out by an Oxford research worker called Orwin showed that others understood the impending changes in farming and the changes that would be wrought on rural employment.

The planned absence of alternative employment opportunities to farming and traditional rural industries could benefit established employers. If high wages and shorter working hours were offered on the countryman's doorstep it would be likely to increase the drift from low paid land-based industry. Where social and geographical uprooting was required the lure of higher paid employment was weaker. That losses from traditional rural employers would occur is supported by the example of events in North and East Scotland as the large pay packets of oil-related developments enticed workers out of farming and a whole range of traditional jobs.

Those who benefited most from the absence of alternative earning opportunities immediately after the Second World War were also those who were well represented in local politics and the control of development was

Figure 5.2 Incomes, employment and rurality. The negative correlation between rurality and unemployment indicates that the more rural the county, the higher the unemployment. By contrast, the positive coefficient between rurality and incomes indicates that incomes fall as counties become more rural.
Source: I. Hodge and M. Whitby

Year	Rurality / Unemployment	Rurality / Income
1976	- 0.6959	0.6252
1977	- 0.6934	0.7308
1978	- 0.7038	0.5133

vested in local authorities. The exclusion of new opportunities for employment would thus help to sustain a low wage rural economy.

However, neither the dissenting member of the Scott Committee nor
Orwin foresaw the scale of the movement of people out of urban areas into
the surrounding countryside to live some 20 to 30 years later. To the majority
of these new residents the countryside was an amenity, not a workplace.
Their green vision was 'ecological-green' – rather than 'agrarian' green and
new employment opportunities could be seen as a threat to their amenity. As
this influx of people grew in number, so did their political power. Their
professional backgrounds were an advantage in lobbying and in pressure
group activities. Rural employment issues were seen as largely inconsequen-
tial, except where amenity was threatened.

The joint effect of the agrarian and ecologically green visions was to
marginalize employment as an issue. Neither group wanted new industry.
Both were prepared in their different ways to accommodate the traditional
village craftsman. The subsequent clashes between the holders of these two
visions were fought largely on amenity and conservation issues which
became the focus for public attention in the debate about the countryside in
the 1970s. This focus in itself further marginalized rural employment as a
political issue.

But in the 1980s we have seen rural employment emerging from the
shadows as an issue of relevance. This has happened for several reasons.
Firstly, the prominence of unemployment and its rapid rise in the early 1980s
at national level made it inevitable that attention would also be focused on its
rural component. Secondly, Britain is experiencing a major experiment in the
field of economic enterprise with the promotion of economic individualism
as the ideal matched by a withdrawal of the state from many fields of activity.
Small business development and a reduction of state control in the land-use
planning system, as we have seen in chapter 4, are both manifestations of this
politically-directed process. Thirdly, agriculture has been singled out as a
suitable case for treatment by the government using analytical tools that were
sharpened on the coal and steel industries. The barricades that Scott had
erected and that had been effectively manned by the European farmers lobby
after entry to the European Community (EC), were breached by the mid
1980s. Finally, in the restructuring of British industry that has been taking
place during a recession and with the application of free market policies,
rural, or at least areas away from the largest conurbations, have been the
beneficiaries of new employment opportunities.

As established values about the rural economy are thus being challenged, a
more individualistic economic philosophy will necessarily lead to a significant reduction of agricultural support. A weakening of the powers of local
government will allow non-agricultural entrepreneurship an easier entry into
the rural economy. As we have already argued in chapters 2 and 4, the twin
props of the Scott philosophy, agricultural support and restrictive planning
control, are being eroded at the same time. Values have been shaken by this
assault. Farmers are asking themselves whether farming is a special case.
Planners are wondering whether the countryside that they have protected
will become victim to the relatively uncontrolled type of development that
occurred in the 1930s.

The boundaries of the new individualism have not yet been firmly drawn.
By mid-1988, it has been rumoured that even National Nature Reserves may
be privatized. Neither have the consequences of these wider changes been
fully elucidated. One result is likely to be the expansion of the breadth of
employment opportunities in the countryside – the rural diversification that
forms the focus of this chapter.

Importantly for rural diversification not only have values about the countryside changed, but values in relation to work and employment have altered too. An economy can be crudely divided into three elements: the market economy producing goods and services which are sold; the state economy producing services (normally) which are often provided on a basis of need; and the informal economy which embraces the black economy, the voluntary sector and the household economy. The values emanating from central government at the moment strongly favour the market economy. Discussion about future work and employment tends to be based on a market element versus a state element. But if neither can offer the prospect of acceptably low levels of unemployment we should not ignore the working of the informal economy. This may have a particularly significant role to play in rural diversification since it is quite a strong tradition in many rural areas.

Experts have attempted to sketch out future scenarios for employment. A first scenario consists of an economy with a developing market element where traditional values of hard work are seen as the prerequisite for material gain and national economic progress. The acceptability of this vision depends on how many are able to benefit from it, and the existence of a significant number of unemployed would be likely to create social tensions which undermine the vision. A second, almost utopian scenario, consists of a society of leisure built upon technological progress, which can emancipate men and women from a long working day and long working life. Questions of attainability and desirability can be asked about this vision of the future. A third scenario consists of full employment, attained by state action as well as

Figure 5.3 The leisure society. Increasing amounts of holiday go hand in hand with a shorter working week, both made possible by technological improvements in productivity.
Source: *Employment Gazette*

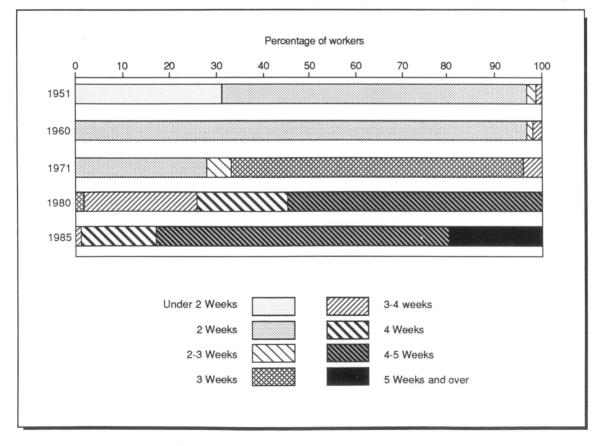

Percentage of workers

Under 2 Weeks
2 Weeks
2-3 Weeks
3 Weeks
3-4 weeks
4 Weeks
4-5 Weeks
5 Weeks and over

private enterprise. This seems improbable in the current political climate and is potentially very costly for the state. A final scenario recognizes that work extends beyond paid employment and that the informal economy may offer opportunities where the formal economy has failed. In this sphere, rural areas might have an edge.

Some of the new residents of the countryside may hold views that would lead them to favour the scenario in which the informal economy had acquired greater prominence. Involvement with village organizations, self-help services, reciprocated skills and vegetable growing are all manifestations of the informal economy which exist in rural areas. Although some of the associated values were in evidence in the traditional village, they have often been enthusiastically adopted by incomers. The informal economy of rural areas has thus received the blessing and the involvement of many new residents of the countryside. The many migrants to remote rural areas in the late 1960s and early 1970s were often the harbingers of these values. Whilst many of these, including hippies and ecofreaks, have been rendered respectable by the passage of time, their attitudes to work and employment are somewhat different to those of the recipient communities. Furthermore, a more acceptable vision of the informal economy has entered the rural economy on the coat-tails of early retired people and green tinged yuppies.

Where does the traditional worker fit into this scenario? Weaned on the work ethic and often finding it reinforced by his employer's behaviour, these new values are somewhat at variance with his view of the world. But as habitat diversity creates new niches, so social diversity creates new employment opportunities. The demand for specialist skills like stonemasonry is likely to be revived and the less skilled may still be able to look for new opportunities stimulated by the imported wealth of these new residents.

Figure 5.4 'It was originally built as a labourer's cottage.'
Source: Reproduced by permission of Norman Thelwell and Methuen

The future of the countryside and the future for rural employment will be shaped by the values of different groups. The values that will be instrumental in determining future outcomes relate both to 'countryside' and to 'employment'. Values associated with both are in a state of flux. The absence of a unitary view is clear. Different values are likely to dominate in different geographical areas. Where farming remains a major industry and diversification opportunities are limited, traditional agricultural values will undoubtedly continue to hold sway. In other areas, a new entrepreneurial spirit may provide an infusion of jobs such as has occurred in many small towns particularly in lowland England. But in yet other areas the entrepreneurial values will be dampened, if not extinguished by conservation values which protect the green image of the village. The prevailing situation in most areas will be one in which competing values coexist, at times uneasily, to provide the context for work and employment in the future countryside. We will be able to distinguish rural areas by the attitudes to the countryside and to employment that prevail in them, as well as by their physical characteristics.

Structural changes and the rural economy

Perceptions and values of and in the countryside are therefore changing. But how does this relate to changes in the physical and economic structure of rural areas? The rural economy of the UK in the late 1980s is an economy in transition. If major structural changes are taking place, there is, however, a remarkable shortage of published data on aggregate employment change in rural areas. As the profile of rural employment has been raised, those who study such matters have been forced to use illustrative rather than comprehensive data. Consequently, there have been differing interpretations of the crucial changes that are occurring.

It is most important not to be lulled into complacency by an aggregate picture of relative prosperity in rural areas with favourable economic trends. Firstly, wage rates are low. Secondly, there are major variations in employment opportunities and unemployment rates from one area to another. For example, in the summer of 1986, eight out of the twelve worst unemployment levels in the country were to be found in predominantly rural areas. Thirdly, attitudes of young people in the countryside show evidence of frustrated ambitions and restricted employment opportunities.

Attempts have been made to distinguish between a southern prosperous rural region and a peripheral impoverished rural region. This would appear to be an oversimplification as prosperous rural areas can be found as far north as Speyside and as far west as Cornwall and Mid Wales. Clearly here again we need to be careful about too simple a definition of 'rurality'.

A final cautionary observation should be the recognition that many of the skills that exist in traditional rural employment are not readily transferable to the profitable and prospering new industry that is infiltrating the rural economy. Instead, a dual economy can be suggested. The traditional 'rural skills' sector may become more impoverished as state support is reduced and the expanding new or 'incoming' industries may provide few opportunities for the traditional community.

There is a marked contrast between those who view the rural economy from the 'inside', who see a decline in job opportunities in traditional industries and services for rural people, and those who view the rural economy from the 'outside' as a relatively prosperous part of the national economy that has experienced buoyancy and growth with an increased share of manufacturing and service employment. Proponents of this 'inside' view have argued that 'rural labour markets are at the core of the problem of rural depopulation in developed countries'. In stark contrast, those who adopt an 'outside' perspective see growth where others see decline. They have argued that the industrial city which grew up in the 19th and early 20th century is proving a transient phenomenon and that manufacturing employment is prospering in rural areas.

In order to explain these differences in the 'inside' and 'outside' views of the rural economy we need to explore the reasoning behind these claims. It is convenient to do this by considering the different roles of primary, secondary and tertiary sectors of the economy in rural employment change.

Declining primary employment

Over most of the present century labour shedding from traditional rural land-using activities has been associated with out-migration, particularly from remote rural areas. The huge decline in the agricultural workforce and the long-term population decline that has occurred in many rural areas in the developed western world reflect major changes in primary industries in these countries. These changes also have a detrimental effect on the numbers employed in providing goods and services to the primary industry.

Figure 5.5 Labour and mechanization. Scything teams began to be replaced by mechanical reapers after 1870, reducing the workforce needed to bring in a hectare of wheat by 75 per cent. Mechanization has continued to replace labour through this century, but at an increasing rate since the last war.
Source: MERL

The loss of workers has not necessarily come about through the decline of primary industry, but often through reorganization. In farming, forestry and mineral extraction there have been major labour productivity gains by workers which reflect the substitution of capital equipment for labour. But industrial reorganization can still lead to job losses. For instance, between 1946 and the present there has been an average annual loss from the farming industry of 2 to 3 per cent of the workforce. This is about one worker per minute in Europe over the previous 20 years.

The agricultural crisis of the mid 1980s has accelerated the loss of employment in the primary sector and had inevitable repercussions on ancillary industries. The immediate effect has been to increase markedly the loss of full-time hired workers from farming. This process will make more and more farms reliant solely on family labour. At a time of high unemploy-

ment in the economy at large there is also a possibility that family members might be absorbed into the family farm without there really being any work for them. It is possible that this reservoir of underemployed personnel could aid the process of diversification, but equally it might create pressures for increased output of conventional products.

But the association between rural labour markets and population decline has been broken by reverse migration. This phenomenon has been observable for at least two decades around larger urban centres, but in the decade 1971 to 1981, reverse migration was also strongly in evidence in remoter rural areas. However, reverse migration does not necessarily indicate an expansion of employment opportunities in rural areas. The migrants may be retired or may commute outside their home area on a daily or weekly basis. This is not an exclusively British phenomenon. There is evidence from other parts of Europe and from North America of a similar process. It is generally

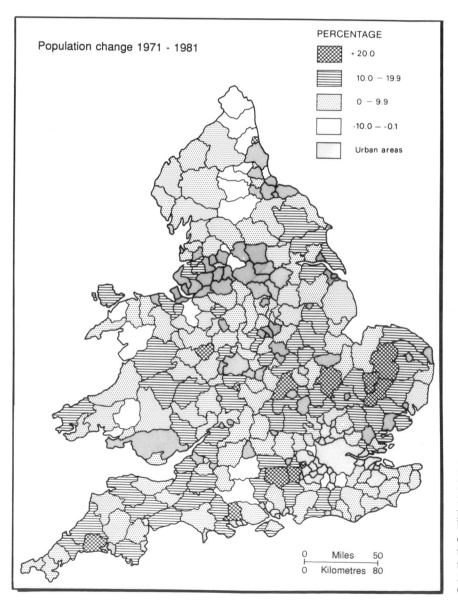

Figure 5.6 Rural repopulation. Population growth in rural districts has been prevalent for some time, but the rate was higher during the 1970s than the 1960s, especially in the remoter areas. *Source*: P. Cloke and G. Edwards

assumed that this process will continue into the 1990s. However, there are also signs that some inner city areas are experiencing their first population gains for many decades. Although these gains are small, the process of inner urban population revival should be looked at closely. The countryside is not necessarily to everyone's taste as a place to live or work.

The new residents of the countryside have often not brought employment opportunities with them. Indeed, their attitude may be antagonistic to certain types of development. But in time these new residents, even if they are not economically active in their home area, are likely to create opportunities for work. They too will need public and private services some of which must be locally provided. Their affluence may be the salvation of local tradesmen. The social impact of these incomers has both positive and negative aspects, as we discussed in *The Changing Countryside*. This is equally true of their impact on employment.

Changing secondary employment

In the past, much manufacturing industry has existed in rural areas. The spread of this industry was by no means uniform and particular activities were concentrated geographically. This was the case with the woollen industry in Norfolk and Suffolk or metal working in the Weald or the Black Country. Individual units of production were often small, but the net effect of this employment on the rural economy was considerable. In the early years of the industrial revolution many larger units were set up in rural areas, factories like Robert Owen's New Lanark or Samuel Greg's mills at Styal in Cheshire, are still in a rural setting today. But in the late 19th century manufacturing employment was drawn into the large towns and cities, and for many decades this process of concentration weakened the residual manufacturing activities of rural areas.

However, the process of concentration of manufacturing has ended. Work carried out at Cambridge on manufacturing employment change, backed up by other research indicates a growth in manufacturing in rural areas and a

Figure 5.7 Nineteenth-century manufacturing in the countryside. Robert Owen's reforming vision of decent conditions of life for factory hands found expression in the building of his own mills in the rural setting of New Lanark.
Source: I. Donnachie

decline in metropolitan areas. This work has contributed much to the 'outside' view of the rural economy. The changes in manufacturing employment are explained by reference to four factors. The industrial structure is important in that some regions are burdened by declining manufacturing industries which have been shedding labour. Urban structure has also been an important influence on change in manufacturing employment. Metropolitan and conurban areas have shed large numbers of manufacturing workers even when the industrial structure is taken into account. The opportunities of new sites for expansion in small towns and rural areas may be a major advantage to them. The size structure of firms is also an important variable. The growth of manufacturing employment has been greatest in areas where the average size of existing firms is small. A final influence on manufacturing employment change is regional policy.

Where industrial structure, urban structure and size structure of firms are all adverse, the loss of manufacturing jobs has been greatest. The areas of greatest growth throughout have been rural, albeit rather more broadly defined than in some studies (see figure 5.8).

The rural data from the Cambridge research has been analysed in some detail. The researchers considered rural manufacturing employment change in the context of different levels of rurality and also by dividing the UK into southern rural region and a peripheral rural region. Although the levels of manufacturing employment increase were greater in the southern rural region, the same gradation from rural growth to metropolitan decline was discernible both in the north and the south. In the peripheral rural region the growth in manufacturing was greatest in the most rural districts although this relationship was not discernible in the southern rural region.

It remains extremely difficult to generalize about rural manufacturing employment. In some areas, the 'inside' view may genuinely prevail. In others the 'outside' picture of employment growth is likely to be more accurate. If the employment base is very small then a small numerical increase can produce a large percentage gain and this is undoubtedly a cause of the high rural figures. We should not, though, dismiss its significance. Although the precise chemistry which leads to new manufacturing opportunities in some areas and not others may be unknown, the net effect on the countryside must be to broaden the range of employment opportunity and stimulate further indirect employment.

The development of service employment

We know less about service employment than about primary or secondary employment. Yet nationally, service employment accounts for over half the workforce. It is a major contributor to national well-being and many consider that an expansion of the service sector is a vital ingredient in national economic recovery.

Our knowledge of the service sector in the rural economy is dominated not by evidence of employment opportunity, but of consumer service decline. We think of shop and garage closures, the decline of public services like transport or the development of ephemeral and menial tourist jobs. These represent the blacker side of service industry change. That such changes have happened is beyond doubt, but there is also another side to service employment change.

Services can be categorized in many ways. A useful starting point is to make a distinction between consumer services and producer services. Consumer services are likely to be provided in a way that reflects the distribution, wealth and mobility of the population. Producer services are services to

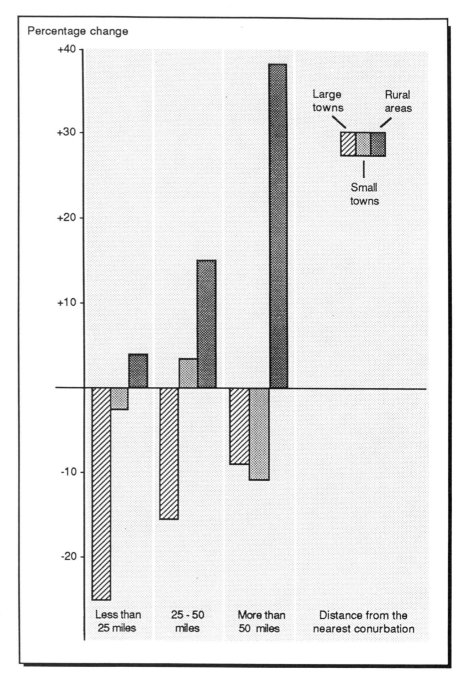

Figure 5.8 Rurality and manufacturing growth. Manufacturing growth has been most evident in the deeper countryside, but it does not follow that wages will be higher or unemployment lower, as we indicated in figure 5.2.
Source: S. Fothergill and G. Gudgeon

industry which need not be too closely tied geographically to the industries that are serviced. These services include such areas as accounting or research and development.

The pattern of change in employment relating to services is complex. There has been a tendency for growth in services to be concentrated in those areas of the south and east away from metropolitan areas – our *golden belt* of chapter 4. In addition, however, there has been rapid growth in some parts of the North, particularly in some rural areas of Scotland as a result of oil-

related developments. As is the case in manufacturing industry, the major conurbations have gained least. However, the southward shift in research and development services and related types of service will tend to aggravate the north-south divide.

Service employment can consist of higher level skilled work or lower level unskilled work. The disdainful attitude shown by traditional rural residents to service employment may reflect their experience of services as lower level, unskilled, low paid work. The movement of high level services in rural areas may be a measurable fact, but it may not represent a broadening of opportunities for traditional rural residents.

To consider rural employment change through green tinted spectacles and to look only at change in employment in agriculture and other primary industries and in supporting service industries is a dangerous oversimplification. Equally it is foolishly optimistic to believe that manufacturing and service employment growth will be the salvation of rural areas. The changes are not felt with equal impact everywhere, but the historic processes of concentration into the industrial city are being reversed in the post-industrial age.

Exactly why these processes of change should be occurring can be explained in various ways. Although the urban decline can be attributed to the old manufacturing industries, there seem to be other factors too. The buoyancy of a small firm economy and physical opportunities for expansion are also considered to be contributory factors to rural and small town expansion. Large cities as well as large firms seem to exert a dampening influence on employment growth.

An alternative critical perspective interprets the flight of industry to peripheral areas as a mature phase in capitalist development, enabling the exploitation of weakly unionized, lowly paid workers in areas where

Figure 5.9 Changing fortunes in service provision. A capacity to meet most of the basic consumer services from one or more retail outlets was once a commonplace aspect of the village, as in this turn of the century example from Berkshire, which supplied groceries, bread, hardware, drapery, clothing, medicines, fuels etc. Today most villagers would have to travel to the nearest town to fulfil a similar range of needs. *Source*: MERL

alternative earning opportunities are limited. Many new jobs created in services or manufacturing are part-time, largely unskilled, female work opportunities. The establishment of branch plants in these areas, often mopping up regional policy aid, suggests that the restructuring idea may have some value in explaining what is happening. If the pattern of employment change was based on the indigenous development of autonomous firms, the restructuring argument could be more easily challenged.

No simple explanation of employment change in rural areas is therefore sufficient. Structural change is in evidence. The nature of these structural changes can be observed. The reasons for these changes are debated, but it is the processes that must be understood if we are to predict the future pattern of countryside employment. Those ideologically rooted into the traditional rural economy look with some pessimism at the decay of the traditional sector and the decline of employment opportunities. But the view from the 'outside' is different from those that see the countryside as an eligible site for development and an appropriate environment in which to locate new manufacturing and service firms.

In the midst of this structural change in rural areas, then, fuelled by both changing agricultural and land-use planning policies, we turn to our third theme of this chapter. We are now concerned to see how government policies, including those for agriculture and planning, are creating a climate for new employment opportunities in the countryside.

A context for rural employment policies

The phrase 'rural employment policy' suggests a comprehensive set of policy mechanisms, the objectives of which would be to benefit rural job prospects. Nothing could be further from the truth in the context of Britain, for employment has been a minor issue in rural policy at a national level. Those policies which have had impacts on rural employment have usually been limited in their operation to particular industries or particular geographical areas. The policies within the Development Board for Rural Wales and the Highlands and Islands Development Board in Scotland came closest to being comprehensive. We should not consider it in any way surprising that policies should be limited by geographical areas. Employment prospects and unemployment rates vary markedly and it would be absurd to have parallel policies in mid-Wales and mid-Suffolk. However, the extent to which exchequer support has been channelled largely into the primary sector of the rural economy, particularly agriculture, is something we should recognise as a major distorting influence on rural employment.

In briefly reviewing the policy instruments that have influenced rural employment since the last war, we can see at once that they constitute the rather haphazardly laid foundations on which recent initiatives in rural diversification have been developed.

The main source of policy support to rural areas clearly has been through the agriculture industry. In the mid 1980s, as we noted in chapter 2, the farm sector in Britain was receiving annually about £1,800 million of exchequer support overall. There were additional costs to the consumer too. The comparable cost of the Council for Small Industries in Rural Areas (CoSIRA), the key rural development agency, was about £20 million. To this can be added the cost of the development boards in rural Wales (about £10 million) and the Highlands and Islands of Scotland (about £30 million). A crude estimate would suggest that on average each farmer in Britain was receiving something similar to the national average wage in subsidy from the state in the mid-1980s, whilst the average manufacturing of service industry worker was receiving only the small change from the national coffers.

The bias of Exchequer support towards the farming industry was rooted in the post war policy proposals and the ideology fostered by the Scott Report. That commitment to agriculture was founded on the assumption that a populous countryside was a consequence of agricultural support. Instead, what has happened has been a continued decline of agricultural labour. In an area like Norfolk, a major beneficiary of the support policies, employment in agriculture fell to about 25 per cent of its post-war level by 1980. The decline has been less marked in areas with a larger proportion of small farms and a greater significance of livestock production, but is nonetheless still markedly downward.

The exact extent to which agricultural support policies have influenced employment change in rural areas is debatable. It is not inconceivable that a combination of stable support prices and production grants for labour saving technology could have led to a more highly capitalized industry with less labour than would have been present in a less distorted market. The decline in the workforce must have had consequences in the demand for consumer services. However, the demand for producer services would be enhanced by a prosperous industry and would create employment amongst the suppliers of farm inputs and processors of farm outputs. But many inputs come from highly capital intensive industries like fertilizer plants, or urban based factories producing farm machinery. The local variety of agricultural machines, which was evident up to the last war, has been replaced by mass manufactured machines, often made overseas. Thus the countryside is unlikely to be the beneficiary of any employment gains that have occurred as a result of the massive capitalization in agriculture.

As we explained in chapter 2 price policy support under the Common Agricultural Policy (CAP), has tended to support larger producers to a greater extent. As price policy has been the major element of domestic policy before and after entry to the EC, the concentration of income support to larger farmers has been inevitable. Development grants to aid smaller farmers have not in general been widely taken up. The Small Farm Business Management Scheme of the late 1960s is a typical example of such a grant with a low take-up rate. The European-based Farm and Horticultural Development Scheme of the early 1980s was potentially more advantageous, but the British government chose to ignore the fact that it was designed for one or two man units and made all sizes of farm potentially eligible. Inevitably it was the large development schemes on the large farms that took much of the grants, but some remoter farms situated in what are termed Less Favoured Areas did take up the scheme, not least because of the additional benefits to these areas. However, since the scheme was for the purposes of modernization, it also tended to be labour shedding.

In the hills and uplands, Less Favoured Areas policy has paid some lip service to maintaining at least the farm population. The subsidies, though, were given for sheep and cattle, not shepherds and cowmen. There has been mounting criticism of the ineffectiveness of conventional hill and upland support policies in retaining people and benefiting the environment. Thus, even in areas where depopulation has been greatest, agricultural policy has been of limited benefit to employment.

Nevertheless the net effect of the whole array of these agricultural policies is extremely difficult to assess. Is it better to retain fewer people in relative wealth rather than a more populous and impoverished agricultural work-force? The critics have argued that there are alternative courses of action that could still have been wealth creating. Some of them would also have undoubtedly benefited the environment and employment in rural areas to a much greater extent.

As we have seen in chapter 3 support has also been dispensed to the forestry industry which has inevitably influenced employment in both private and public sectors. The private sector has received production grants and tax concessions. The consequence is likely to have been a small amount of additional job creation which has reached locally significant levels in particular areas. The Forestry Commission as an enterprise in its own right has certainly created jobs, often in areas of relatively high unemployment. In the inter-war period job creation was actively pursued as a policy instrument. Social policy in the forest sector has largely evaporated under the cost-conscious eye of the Treasury since 1972. There has been an accelerating trend of labour shedding as machines have replaced workers, despite the increasing importance of job creation as a forest policy objective. Much forestry work is now carried out by itinerant gangs of workers which does little to enhance the stability of the local employment base. Nonetheless, there are likely to be indirect employment effects in timber processing which must also be recognized.

Occasionally other primary industries have been affected by specific policy instruments. The development of hydro-electric power in the post-war Highlands created many jobs during the construction phase. National energy policy has positively affected rural coal mining jobs in the vale of Belvoir (although doing little to off-set other pit closures), as well as oil-related jobs in northern Scotland. Often, however, non-agricultural primary industry has depended on the general instruments of regional policy in the area concerned. What stands out in any examination of the impact of these policies on the rural economy is the fragile base on which prosperity is often founded.

As well as agriculture and forestry policies, regional policy instruments have at various times influenced employment creation in rural areas. The focus of regional policy has always been the decaying urban industrial regions. Recognition that unemployment problems extended into rural areas led to a widening of its application to include rural areas in the 1960s and 1970s. Initially, the support given focused on manufacturing employment creation, but from the mid-1970s onwards a range of services has also been included. An ambivalent attitude to service employment remains with some services receiving mandatory grants and others being considered on a discretionary basis. At their fullest extent in the late-1970s, varying levels of regional aid were given to large areas of rural Britain including virtually all of the United Kingdom north of the Humber and Mersey, west of the Exe and most of Wales and Scotland.

Whatever the general state of regional policy measures over the years, the two development boards in rural areas have remained as significant influences on employment creation in the Highlands and Islands of Scotland and in mid-Wales. Both were initiated in the mid-1960s. The Highlands and Islands Development Board was given a particularly broad remit embracing grant-aid, loan finance and research with all sectors of the economy able to receive assistance. A further development board had been established in the North Pennines in England with a much more explicitly rural land-using remit under the 1967 Agriculture Act, but this early experiment disappeared on a change of government before any merits of its existence could really be tested.

The development boards that survived have been instrumental in creating jobs in their respective areas. In their early years, when the focus was very much on job creation, both were criticized for ignoring social and community development. Prior to the more depressed economic climate in these remoter

Figure 5.10
Employment in the Highlands and Islands.
The Highlands and Islands Development Board have done much to develop commercial salmon fisheries as one means of stimulating employment in remoter rural areas.
Source: Highlands and Islands Development Board

areas of recent years, branch plants or external developers could be attracted in. This led to accusations of colonial style development in the two areas from some quarters. Development boards have always had problems in having to justify themselves to two very different groups: government bureaucrats and the often distinctive local communities of the areas in which they operate. But they have tended to respond in part to these criticisms from the indigenous population through the development of community cooperatives in many remoter areas of the Highlands and Islands. Unfortunately, central government tends to measure policy success by cost per job created and community benefits that cannot be easily measured attract less attention.

The Rural Development Commission (now including CoSIRA) and equivalent bodies in Scotland and Wales, constitute the final element of central government policy that has influenced rural employment. The importance of these bodies is that their remit has been exclusively rural. Their role reflects fairly closely national attitudes to the non-primary element of the rural economy.

The Development Commission as it was called when it was set up in 1909, has acquired and shed various responsibilities over the years. It has undergone various identity crises and, to extend the medical analogy, could have been diagnosed as schizophrenic until relatively recently. The principal problem has been whether or not its remit was limited to traditional rural crafts or whether it extended to the whole expanding range of industries that operate in the countryside. If it was just a craft development agency it would be a different organization to that which sought to act as a regional policy agency in a rural context. The search for an identity has produced a broad ranging agency which is perhaps stronger today than at any time in the last forty years.

The principal role of CoSIRA, which merged with the Development Commission in 1988 to form the Rural Development Commission, is to develop small scale factories and workshops in rural areas. Historically, the support was exclusively for manufacturing industry, but CoSIRA has edged cautiously into supporting certain types of service industry. In addition to the provision of sites, a wide range of business services is offered including advice, training and credit. Their activities have been concentrated in four

Figure 5.11 'Restoring' rural employment. Non-agricultural employment in the countryside has long been supported by CoSIRA and the Development Commission.
Source: CoSIRA

main areas: the South West, the Welsh Borders, the North and the shire counties of the North Sea coast. Separate arrangements operate in Scotland and Wales.

Although CoSIRA is a rural agency it has not always been able to show that it can have a 'Heineken effect' on rural employment by reaching the parts that other policies cannot reach. Inevitably small-scale factories are located in large villages or market towns and can still be relatively inaccessible to displaced workers from the primary industry. Rural employment opportunities thus depend on access and the new Rural Development Commission, is increasingly finding itself lobbying on issues like rural transport planning and funding experimental schemes to improve mobility. Certainly the role of the Rural Development Commission has long extended beyond factory building and business services. Rural Community Councils, financially supported by the Rural Development Commission and the Welsh Office, are county-based voluntary organizations which have been instrumental in developing pioneering approaches to the solution of rural problems. Where the market or the state sectors have failed, they have been able to activate the informal economy and to catalyse action in community shops, transport provision and a range of other services.

Thus, at the beginning of the 1980s countryside employment was being influenced by a number of different policies and organizations. In terms of government funding, the traditional primary sector, and within this farming and forestry, was the principal beneficiary. The rural economy, though, was losing its identity as more mobile industries relocated themselves in rural areas. Out of these changes CoSIRA and the Development Boards more than ever established themselves as agencies with an important role in the changing countryside. As the economy in the late 1970s lurched into deeper

Figure 5.12 Activating the informal economy. Rural Community Councils have proved an effective means of providing advice at the grass roots level. *Source*: ACRE

recession and the profligacy of agricultural policy was subjected to closer scrutiny, the new decade offered new challenges for the countryside. It is to the response to that challenge and its impact on the rural economy and jobs that we will now turn, as a fourth theme in this chapter.

New initiatives and rural employment: the primary sector

The prospects for rural employment have altered in response to changes in the market economy and changes in the policy environment. The principal changes in rural employment over the last forty years have been the expansion of secondary and tertiary sector employment and decline in the primary sector. The strength of the forces of expansion and decline of the different sectors have been by no means even across the face of the country. This has inevitably created significant disparities of opportunity.

As we saw in chapter 2, the 1980s began with a series of attacks on agricultural policy and its beneficiaries. The instruments of policy were criticized as being politically unacceptable, financially extravagant and damaging to the environment. The criticisms were articulated in many places in many ways. Books of a polemical nature like Marion Shoard's *The Theft of the Countryside*, stirred up a spate of letter writing to newspapers and precipitated a number of television documentaries which were often highly critical of the agricultural industry. Even agricultural economists added their comments on what they saw as the illogicality of the CAP. The result has been a weakening of the farming lobby's power and the development of policies to curb agricultural over-production. It is now widely recognized that restraint in policy expenditure on agriculture will occur and out of this changed and changing policy climate a debate has arisen on the future of the rural economy. Bearing in mind the multiplicity of pressure groups with an interest in rural affairs, we can expect no consensus on the desired result. But out of this debate new policies are emerging, the consequences of which, for the rural economy, have yet to be fully assessed. Undoubtedly, though, they will have a significant impact on rural employment as the economy diversifies.

Almost forty years of agricultural stability and prosperity have been associated with a massive reduction in the farm workforce and the perpetuation of a low wage rural economy. After two or three years of tentative policy reform and stagnating or declining farm incomes, further employment decline has resulted. To the pessimist a policy beneficial to employment seems a forlorn hope. But in the present climate of policy uncertainty, at least rural employment and the functioning of the rural economy as a whole are receiving greater attention than at any time in the recent past.

The British government has championed the cause of CAP reform, but the pace of change is constrained by the political strength of the European agricultural lobby and the way in which European decisions are made. It is ironic that the Ministry of Agriculture, Fisheries and Food (MAFF) should now be actively promoting the cause of rural diversification when the same ministry was so slow to implement the craft and tourist components of the Less Favoured Areas Directive during the late 1970s. However, for a variety of reasons the pace of change is quickening and, as we noted in chapter 1, the 1986 Agriculture Act has broadened the Minister of Agriculture's responsibilities to include 'the economic and social interests of rural areas'. In 1987 MAFF published *Farming UK*, a glossy publication replete with references to the diversification of the rural economy. Cynics might see this as nothing more than a change in public relations strategy. Optimists have argued that the seeds of policy change have been sown.

The ideas contained in *Farming UK* had been launched as the ALURE package in the Spring of 1987. Its appearance had been surrounded by internecine strife and some of the proposals were greeted by disapproving noises from a wide range of critics. A measure which had been devised to rebuild the tarnished image of MAFF did little to achieve this, at least in the short-term, a point that we considered when we laid down the political context for rural change in the Preface to this volume.

The unfavourable reception given to ALURE hinged, at least in part, around the anticipated threat to amenity. The most controversial element was the weakening of MAFF's vetting powers over development involving lower grades of agricultural land, as we discussed in chapter 4. It is difficult to disagree with the logic of this move for it did nothing more than remove powers which were introduced to safeguard agricultural production, hardly a priority in a time of costly surpluses. However, coupled with anti-planning, pro-development sentiments emanating from some sections of the Department of the Environment (DoE) and the overall government philosophy favouring the weakening of the power of the state, the amenity groups foresaw problems. They envisaged England's green and pleasant land pockmarked by sporadic developments, destroying the image of the countryside that they nurtured. They have only been partially mollified by subsequent assurances that countryside and heritage will be safeguarded.

The development of farm woodland constituted a second strand in the ALURE strategy as we discussed in chapter 1. In the early 1980s the Forestry Commission had turned the old Dedication Schemes into a streamlined grant scheme and launched the Broadleaved Woodland Grant Scheme to support its more positive attitude to broadleaved woodland. The ALURE proposals offered an annual payment to farmers for a given period of years for afforesting land. It was hoped that the combined effect of the Forestry Commission grants and the ALURE payments would persuade farmers to take land out of production of surplus products and plant trees instead. This new policy for woodlands is out of tune with the government's philosophy and is founded more on hope than any expectation of success.

Figure 5.13 New policies and proposals. A proliferation of recent policy documents put forward by the government and others has inevitably led to disagreement and conflict.
Source: D. Noton

The employment benefits of farm woodland development are thus likely to be negligible. Silviculture is no more labour intensive than most agricultural systems in the lowlands and rather less labour intensive than many. There may, however, be some scope for diversifying farm and woodland employment and so benefitting the rural estate as a whole.

The third element of the ALURE package was the provision of a range of grants to aid diversification of farm business. The grant for diversification into a specified range of alternatives has been set at 25 per cent of approved expenditure up to a ceiling of £8,750. Eligibility criteria have been established to weed out non-bona fide farmers. Grants are also to be made available for feasibility studies and the marketing of diversified products and rates for these are likely to be set at a higher level.

The ALURE proposals represent a significant shift in modes of thought within MAFF, but the consequences on the ground are unlikely to be dramatic. The financial commitment to diversification is miniscule when set alongside the total costs of agricultural support. Only £3 million has been put aside for diversification grants in 1988 to set against an anticipated direct policy cost of some £1,800 million. Furthermore, the diversification scheme is only a national measure. This is quite unlike the EC wide Environmentally Sensitive Areas measure, a conservation approach which we will examine in some detail in chapter 7 and is therefore vulnerable to being cut by budget trimming in the event of a financial crisis.

Farming and landowning pressure groups have generally welcomed the ALURE proposals, but have predictably sought higher levels of grant than those initially offered. The conversion of MAFF to the cause of diversification was preceded by the conversion of representative pressure groups. *The Gretton Report* from the County Landowners' Association in 1985 has been described as a radical document. It advocated diversification on the farm and in the rural economy at large, with a much stronger role for the Development Commission and CoSIRA. The National Farmers Union's *Way Forward* was much more realistic in its assessment of the climate of policy-making than earlier policy statements. These shifts in policy at the headquarters of these organizations towards a new realism took place at too fast a pace for many members in the shires. A visit to a branch meeting would be a powerful reminder of the outmoded values held in the ranks. Their way forward might best be described as backwards into an earlier agrarian economy through rose tinted spectacles. Nonetheless, the changes at the centre of these organizations is likely to filter into the provinces. Policy and pressure group activity can aid the transformation of the rural economy, but ultimately it is the farmer or the landowner whose actions will determine the extent of change.

In the search for solutions, as well as in the diagnosis of employment problems, there is a contrast between the views of those inside the traditional community (physically or spiritually) and those outside it. To those inside the traditional community the vision of the future is a more sombre one, characterized by income difficulties and bankruptcies. A similar vision is held by the supplier of farm inputs or the processor of agricultural products. For all of them the future will be difficult; jobs will be hard to find; profit margins low. To many of them there is a feeling that what ought to be gratitude from the urban population for food, has been transformed into ingratitude because of what they perceive as minor environmental misdemeanours. Policy makers for the traditional sector have realized that there has been a sea-change in attitudes towards farming. They have been instrumental in trying to encourage diversification. But for a number of reasons we cannot expect rapid diversification of the agrarian economy.

Farmers and the farming press have traditionally been rather dismissive of diversification, seeing it as an opportunity for the physically infirm, the agriculturally inept or the socially eccentric. In the years of policy support for a relatively narrow range of products, the uncertainty of the free market was undesirable to most farmers. These attitudes may be less relevant at a time of declining support for the traditional products. Some entrepreneurial farmers, sensing that the industry's good fortune could not last, moved into alternatives in a highly businesslike manner. These individuals may be the harbingers of a new age, but that new age is likely to take time in coming.

The self-evident entrepreneurial skills required of the diversifying farmer make it appeal to those who are market oriented rather than production oriented. Those who measure their achievement in tonnes per hectare or litres per cow are still fixed into a production orientation. They tend to have an affection for the livestock market rather than the marketing philosophy. The new skills required of the diversifying farmer have been taught in only a few agricultural education establishments until recently. In some cases entrepreneurial skills that could be directed into profitable alternatives are stifled by landlords' conservatism.

Four main areas of potential diversification do, however, exist (see figure 5.14). Tourism and recreation developments have been one of the most widespread forms of this, but there is now real concern in some areas that oversupply might reduce the profitability of these enterprises. Moreover, it is often difficult to generate direct incomes from them, a point we develop in the next chapter. Certainly competition is likely to sharpen the marketing skills of established and new operators. Adding value to farm products represents a second type of diversification. This can be achieved by marketing products differently as in pick-your-own enterprises or by processing as in the manufacture of meat products. The potency of rural images is much in evidence in food marketing. If the product can be consistently produced, as well as effectively marketed, there may be a niche for such enterprises in spite of the dominance of multiple grocery chains. A third type of diversification is into alternative crops and livestock, often unsupported by policy instruments. Finally, farm resources can be re-allocated to alternative uses by building conversions or silvicultural developments. This may take the land out of farming hands, but the new enterprises may be, in some cases, carried on by the farmer.

The challenge of on-farm diversification is to be able to redeploy farm resources to satisfy better the needs of a changing market place. This requires an understanding of the available resources rather different to that required for conventional farming. Farmers must be able to assess the benefits of a particular location for a particular activity. There are likely to be major geographical variations in opportunities leaving some farmers with several opportunities and many with none. Diversification on farms is not a panacea for the present crisis in the agricultural industry. It is one way in which farmers might find their way out of the mire. But the beneficiaries are unlikely to be those most in need of help. Entrepreneurship favours the strong and if policy fails to protect the weak, an accelerated turnover of land can be anticipated.

There are benefits of on-farm diversification. Incomes may be enhanced and jobs may be created or saved. Seasonal work may be created which might be better than no work at all. The presence of visitors on farms might encourage a caring attitude to the countryside. And, importantly, the isolated position of conventional farming businesses will begin to break down. The grant-aid, though, is only directed at bona fide farmers and perpetuates the inequality between farmers and the rest of the rural community. A truly

TOURIST AND RECREATION	VALUE - ADDED
TOURISM: Self catering Serviced accommodation Activity holidays	**BY MARKETING:** Pick your own Home delivered products Farm gate sales
RECREATION: Farm visitor centre Farm museum Restaurant / tea room	**BY PROCESSING:** Meat products - patés etc Horticultural products to jam Farmhouse cider Farmhouse cheese
UNCONVENTIONAL PRODUCTS	**ANCILLARY RESOURCES**
LIVESTOCK: Sheep for milk Goats Snails	**BUILDINGS:** For craft units For homes For tourist accommodation
CROPS: Borage Evening primrose Organic crops	**WOODLANDS** For timber For game For tax avoidance **WETLANDS:** For lakes For game

Figure 5.14 Potential farm diversification. An increasing range of alternative farm enterprises is being developed, some of which are heavily dependent on location and the skills of the farmer.
Source: B. Slee

diversified rural economy would require no elaborate specification of agricultural credentials as a prerequisite for policy assistance.

Policies relating to manufacturing industry

The decline in the significance of regional policy in rural areas, with the exception of the continued activities of the development boards, has refocused attention onto the Rural Development Commission as the key agent in rural employment policy. Following a review of its functions in 1980 the Development Commission was able to draw up its own rules on priority areas for action. This resulted in the designation of Rural Development Areas and within these areas Rural Development Programmes were to be determined by the Rural Development Commission, the county and district councils and other relevant bodies. The means by which the Rural Development Commission assists the development of industry are largely unchanged, although there is increased focus on coordinated action. The ACCORD scheme, launched in 1987 by CoSIRA, offers opportunities for higher levels of funding for certain projects in Rural Development Areas. This is an experimental scheme for a fixed period of time to encourage additional private sector investment in job creating projects by offering up to 40 per cent of what are euphemistically termed 'grants'. Scrutiny of the small print reveals that unlike agricultural grants, 'the Commission will look to recover its grant from projects that have proved successful'. Furthermore, applications for these grants are competitive with only a fixed amount of funding available.

Through this and other schemes, the involvement of county and district councils in employment policy has increased in the last decade. At the time of initiation of structure plans the counties were seen as the key agents in the local authority sector. However, the district councils have acquired additional planning responsibilities and many are now active in the provision of industrial estates. The districts have lobbied hard for additional resources for Rural Development Programmes and advocated their extension to include a wider range of economic activity.

The policy framework has historically been set by the public sector agencies who have also been principal agents in implementing schemes of employment creation. However, the role of the private sector is increasing. Local enterprise agencies exist in many areas, funded largely but not exclusively by the private sector. They, like the Rural Development Commission, will advise and guide individuals in the private sector. They also have received the blessing of the present government and fit in with its policy of privatization. In spite of the insistence of many observers on coordination as the key to rural employment policy, an extension of private sector business advising agencies could undermine the coordination process that is being advocated.

We should not of course forget that the success of any policy depends on the attraction of entrepreneurs who would not otherwise have developed their businesses in the recipient regions. If the firms were prepared to move of their own accord the policy lures are superfluous. The Cambridge research work that we referred to earlier in this chapter, might lead us to conclude that there was little need for a policy for rural manufacturing industry. However, local conditions are highly variable and the initiation of programmes for development targetted on the problem areas offers some hope for the diversification of the rural economy.

Policies relating to services

The national significance of the service sector, in terms of both size and growth, is not reflected in policies to promote service employment. This is a reflection of negative attitudes to service employment and a belief that the service sector is not in need of supporting policies. However, some types of service employment are covered by regional assistance and the Rural Development Commission will give grants for certain services in Rural Development Areas. The lack of compatibility over services supported indicates that the process of coordination is as yet incomplete.

One service industry, tourism, has long been the beneficiary of government support. At times the grant-aid has been targetted on areas in receipt of regional assistance, but this is not currently the case. Grant-aid for tourist projects is split between major projects and small business grants. Grants are discretionary and are given for specified types of investment. There is thus a major difference between these grants and those given by CoSIRA or MAFF for barn conversions for tourist accommodation. The attempt of the government and government agencies to shift negative attitudes can be seen in the *Pleasure Leisure and Jobs* report from the Cabinet Office and the tape/slide presentations developed by the English Tourist Board to present a glamorous and favourable image of the tourist sector.

The tourist industry will continue to generate contradictory images of expansion and decline, seediness and glamour, profitability and insolvency. These conflicting images reflect the variety of the tourist sector, at least some

of which can be found in rural areas. One advantage that rural areas have is that they are not burdened by large decaying resorts. The base level of tourism is often low and the emergent structure of tourism is more likely to be compatible with the needs of the late 1980s and beyond. However, in the farm sector of the rural economy it would be a mistake to see farm tourism as a panacea for the industry's problems and an oversupply of tourist services is likely to emerge in some areas. Different sectors of the industry generate different secondary job opportunities and it may be desirable to target grant to aid where the effects are greatest. It is, however, equally desirable not to ignore market signals and trends. We return to the issue of generating incomes from rural leisure in the next chapter.

The outside view of the prospects for employment in the rural economy are founded to a much greater extent on the prospects for manufacturing industry and services. The aggregate picture for these sectors is one of some prosperity. Consequently, agencies like the Rural Development Commission can sound almost complacent as the number of rural jobless falls in spite of the continual shedding of agricultural labour. But as we have seen, the employment change is as much a product of firms voluntarily leaving urban locations as of firms being lured to rural regions by government policies.

Images of the future

Many would say that rural diversification to date has been hampered by a lack of policy integration and coordination. In some instances, as is the case of rural employment, there has been a lack of specific policies altogether. Some people, as we will see in chapter 8, have suggested that this situation may be improved by the introduction of some sort of 'Ministry of Rural Affairs' with a broad responsibility for rural development. This, however, flies in the face of the modern trend of increased specialization in government ministries and quangos and even in specific fields of rural development. An increased specialization of interests is commonly associated with an increased position of power.

If we are to develop the integration of policies and the diversification of the rural economy we could develop both concepts through 'signposting'. This is a process whereby public, private or voluntary sector agencies help individuals to find appropriate sources of assistance. In a countryside that has suffered as a result of a lack of integrating policies the development of signposting services may be of assistance to some. Effective signposting is not just about knowing which agency provides what level of grant, but also is about finding the personality who can catalyse change. In rural regions the strength of personalities may be a more powerful asset than the amount of grant aid, a line of argument which we take up again in chapter 8.

But even the notion of integration may be a bit of a battleground. On the one hand, the urban refugees of the 1970s and those of an ecologically green tinge may see its potential in procuring environmental harmony. On the other, the promotion of entrepreneurial values and free market economic policy in the late 1980s may cause the integration of policies to work entirely in the service of economic development.

If the entrepreneurial faction wins there is almost unlimited scope for changes in the nature and extent of employment opportunities in the rural economy. We can only speculate as to whether such jobs as might be created would be those of the late twentieth century sweat shop or would consist of satisfying and remunerative employment for the whole range of rural residents.

As the various groups jockey for position and endeavour to sustain and

Figure 5.15 Tourism and the countryside. Many of the more remote hotels, like this one in Cumbria, have come into existence as a result of grant-aid from organizations such as the English Tourist Board.
Source: Countryside Commission

develop their rather different visions, one group, the working class of the traditional community remains largely powerless. The drift from the land will continue. Some new opportunities might appear. Old crafts will be sought by new residents and domestic service in the homes and gardens of the affluent might increase. The primary production skills of farm workers and farmers will be subtly altered as landscape gardening rather than farming takes up their time in Environmentally Sensitive Areas or National Parks. Their preferences in the search for a new integrated countryside are likely to carry little weight.

Certainly we may expect market forces to permeate the rural economy to a greater extent. Policy could have a weaker role in the future and as a result the objective of integration may prove as elusive as ever. In the farm sector, market forces may cause the process of diversification and its extent to vary dramatically from one area to another. However, it is the secondary and tertiary sectors of the economy that we may expect to be most important in creating jobs and diversifying the rural economy. Policy measures may try to plug the gaps created by the geographical variations in opportunity, but there will probably remain significantly lower incomes and limited opportunities in many areas into the foreseeable future.

There is, however, more to the debate about diversifying the rural economy than the issue of policy integration. At the turn of the 1980s a group of development experts visited Britain to observe and comment on strategies and approaches to UK rural development. In addition to their predictable dismay about a failure to develop integrating policies, they also noted the limited extent to which rural development in the UK was rooted in the values and aspirations of rural residents. An alternative focus for the debate about diversifying the rural economy concerns the extent to which externally funded, externally designed, top downwards notions of development are imposed on rural communities and threaten what many would argue were

Figure 5.16 Rural skills revived. Some traditional skills, such as dry stone walling, can be given new life amongst the rural labour force as agricultural employment declines.
Source: Countryside Commission

distinctive and valuable local cultures. Again, this is a matter we pursue in chapter 8.

It has been argued with some conviction that development agency strategies that rely on branch plant developments in rural areas are both an unreliable and inappropriate form of diversifying the rural economy. Yet development agencies, conscious of central government looking over their shoulders at their cost-effectiveness, have inevitably directed their attention to the large firms contemplating a branch plant, rather than the small entrepreneur trying to establish himself in a one-man business. This strategy for economic renewal based on attracting footloose high technology industry into rural areas exposes the sham of the political posturing which purports to revitalize the economy by encouraging small businesses and local initiatives.

There is then, an inherent conflict in diversifying the rural economy between corporatism, into which development agencies so easily lapse, and individualism. In the light of this, can we look forward with any optimism to a unitary rural economy, where the whole community has become a beneficiary of diversification? Can we foresee a state wherein the marriage of the yokel and the yuppie has been consummated? Or will there continue to be antagonisms and skirmishes between these two loosely idemtified groups, whose interests may not converge? In most conflicts the infantry or the innocent bear a substantial burden. A change in the dominant social group creates scope for some social and economic mobility. But a decade hence students of the countryside are likely to be seeking solutions to the problems of residual poverty, low wages and restricted economic opportunities over substantial areas of rural Britain. This is a point we return to in the final chapter.

Figure 5.17 'Another diversification scheme underway then Huw?' Source: Farmers' Weekly

Chapter 6

Increased rural leisure: recreation for all?

The new rural policy era

So far in this volume we have evaluated policy proposals – for agriculture, forestry, land-use planning and diversification – that are to a large extent internal to the rural economy. In chapters 6 and 7 we now turn to an evaluation of policies – for recreation and conservation – that although applied in rural areas, impinge more directly on the population as a whole.

Policies for recreation and tourism in the late 1980s are in some ways more radical than for many other spheres of rural activity. This is not because their impact will necessarily change the visual or economic face of the countryside, but rather because they represent a significant reversal of prevailing policy attitudes of the previous twenty years.

Simply put, since the middle of the 1960s to the mid-1980s both central and local government had been obsessed with the threat of a recreation 'explosion' in the countryside. Public policy concerns at all levels had been to ensure that such a growth in countryside recreation did not lead to an irreversible deterioration in the rural environment that people had come to enjoy. Apart from some attempts at correcting social imbalances in recreation opportunities, policies were therefore concerned with tolerating and in many instances controlling, people's propensities for countryside recreation.

Today, the picture is very different. You will see from chapter 1 that government policies, particularly driven by the Countryside Commission's *Recreation 2000* policy review which has culminated in the 1987 policy document *Policies for Enjoying the Countryside*, are broadly concerned to promote and market rural leisure to as wide a public as possible. The change in policy from a perceived need to control the public, to their wholesale encouragement has happened quite swiftly and has taken place in tandem with a fundamental reappraisal of the primary role of agriculture in rural areas – a reappraisal that underlies all of the issues discussed in this volume.

In this chapter, we evaluate critically this process of policy change in rural leisure. We focus on changes in countryside recreation policy rather than that of rural tourism. These two are always hard to disentangle, but the essential difference for rural areas is that holiday tourism involves a stay away from home of one night or more. Apart from this, rural tourists are essentially recreationists on holiday. We return to the peculiar aspect of tourism, overnight accommodation, when we examine the income potentials of rural leisure later in the chapter. We have already touched on its employment and economy-sustaining potential in the previous chapter.

This evaluation of policy change is undertaken in a number of separate,

Figure 6.1 Riding as rural recreation – in the Peak District. Recent government policies have moved in the direction of encouraging such pursuits in the countryside rather than merely tolerating them. *Source*: Countryside Commission

though related, phases. Firstly, we look at how important rural leisure is and has been over the previous fifteen years, to the public at large. Perceptions (rather than necessarily a full understanding) of the popularity of countryside recreation have been one of the prime determinants of policies over this period, so it will be useful to examine some statistics about recreation participation, what people like to do and also who rural leisure participants actually are. Is countryside recreation indeed recreation for all? Secondly, in the light of this understanding of recreation participants we examine what we will call social policies for rural leisure to see how effective they have been in broadening the base of participation.

Next, we move on to consider a number of policies and characteristics of recreation land-use planning that have prevailed during the 1970s and 1980s and those that currently are being put forward and assess their strengths and weaknesses. We do this particularly in the light of our understanding of recreation behaviour outlined in the first part of this chapter.

From this analysis of policies both past and present, we move on to consider a particularly important issue for rural leisure when it is set in the context of rural change more generally. This is the economic and income potential of recreation and tourism. Finally, we indulge in a more speculative assessment about how policies for rural leisure and the activities of rural leisure will develop to the turn of the century.

The importance of rural leisure

Undertaking trips to rural areas for leisure purposes has never been a universal preoccupation, but it certainly can be very popular. According to the Countryside Commission up to 18 million people may venture out of the towns and into the countryside on a sunny Summer Sunday afternoon.

It will be useful at the start of this chapter to provide you with a summary of the recreation behaviour of the people of England and Wales to give you an impression of both its size and structure and who the recreationists actually are. Importantly, though, we are also going to use this information about observed recreation behaviour to develop a critique of existing recreation policies and the policy proposals of current deliberations on the part of a number of governmental and other organizations.

We must be realistic about the quality of the data on recreation participation and behaviour that currently is available. Much recreation planning, certainly up until the late 1970s was based very much upon assumptions about both the types and extent of recreation activity. The data bases available to planners and policymakers are now much better, but they are still subject to the limitations of relatively small samples attempting to represent the behaviour of the nation as a whole.

It was fashionable during the 1970s to undertake surveys of people's behaviour at recreation sites, and use these as an estimation of overall behaviour. These had severe limitations, however, in informing planners and managers both about non-participation and about behaviour at other, quite different, sites. It was only really with the development of large scale household surveys that our understanding of recreation behaviour improved considerably.

Up until 1977, our countryside recreation household statistics were limited to one or two questions in the General Household Survey, every three or four years and a couple of specifically commissioned studies. Since that time, however, the Countryside Commission has undertaken three large national household surveys in 1977, 1980 and 1984, known as the National Surveys of Countryside Recreation (NSCR). These represent the most comprehensive estimations of a wide variety of characteristics about recreation, and because of this, we will use these three databases in this section, except where otherwise stated.

In absolute terms, according to the 1984 NSCR, the variation in rural recreation trip-making from day to day is quite large, but there are fewer seasonal fluctuations than might be expected. A typical winter weekday saw around 2 million people enjoying the countryside, but a Summer Sunday exhibited a peak of 18 million recreationists – around two fifths of the entire population.

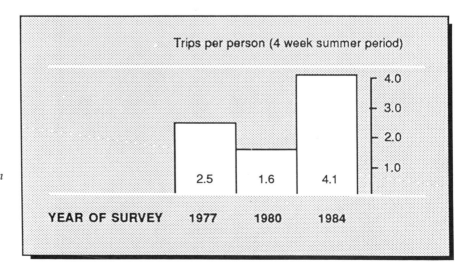

Figure 6.2(a) Recreation participation patterns. The users of the countryside tend to do so regularly
Source: Countryside Commission

Trips per person (4 week summer period)

YEAR OF SURVEY	1977	1980	1984
	2.5	1.6	4.1

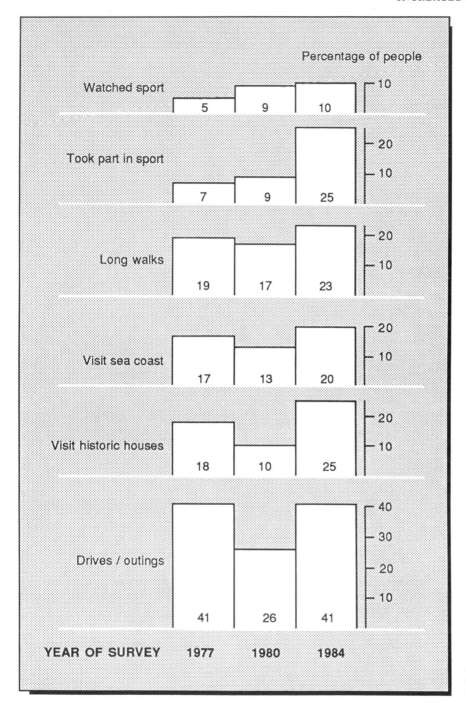

Figure 6.2(b) . . . but the sorts of things visitors do vary in popularity.

The number of trips made by people also varied from one year to the next in the three years of the NSCR. Figure 6.2(a) indicates the variation in the average number of trips per person across the three years of the survey period, for those who were recreation-active in the four weeks prior to each of the surveys. There is no real pattern in this trend, however, since the strongest determinant of these frequencies was, and for countryside

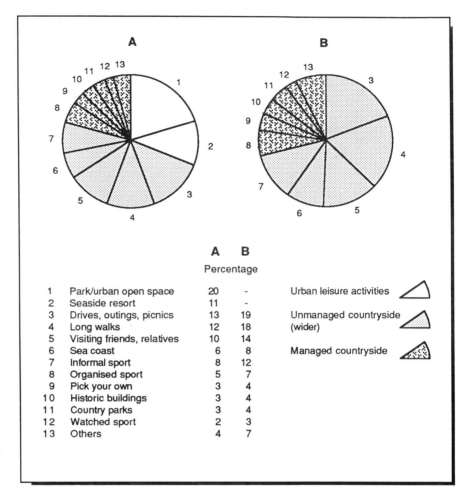

		A	B
		Percentage	
1	Park/urban open space	20	-
2	Seaside resort	11	-
3	Drives, outings, picnics	13	19
4	Long walks	12	18
5	Visiting friends, relatives	10	14
6	Sea coast	6	8
7	Informal sport	8	12
8	Organised sport	5	7
9	Pick your own	3	4
10	Historic buildings	3	4
11	Country parks	3	4
12	Watched sport	2	3
13	Others	4	7

Urban leisure activities

Unmanaged countryside (wider)

Managed countryside

Figure 6.2(c) Out of town visitors are keen to experience the unmanaged countryside

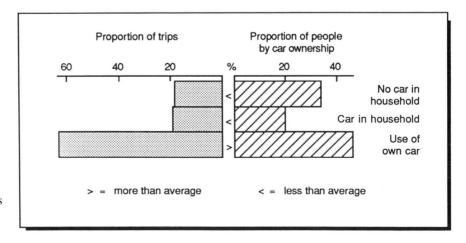

Figure 6.2(d) ... and this invariably means they will need access to a car.

recreation participation always will be, the weather, which was poorer in the summer of 1980 than in the other two years.

You can get some idea of the relative popularity of different types of recreation activity in the countryside from figure 6.2(b). This shows the percentage of recreation-active people who undertook different activities during the four weeks prior to the survey. Again, trends are hard to disentangle because of the weather, but clearly, informal drives and outings in the countryside are consistently the most popular form of countryside recreation. Visiting the sea coast, historic houses and going on long walks are the next most popular, but taking part in active sports in the countryside seems to be on the increase more than other activities, a point that we noted in chapter 1.

For 1984 alone, it is possible to look at the relative popularity of different types of countryside recreation activity in a little more detail. The two pie charts in figure 6.2(c) indicate the relative frequency of use of some eleven different types of countryside pursuit, compared to that of urban outdoor recreation, and visiting seaside resorts. It is interesting to see the extent to which people seem to prefer recreation in the unmanaged rather than the managed countryside, a characteristic that we shall return to when we consider the nature of recreation land-use policies.

It is also useful for us to have some idea of the relationship of car ownership to recreation participation, since this will help us later in our analysis of recreation social policies, when we look specifically at the issue of recreation transport. Figure 6.2(d) examines this relationship for 1984 in simple terms. The figure indicates that where there is no car in the household, the number of countryside recreation trips will be much lower, proportionately, than where there is full access to a car. This picture has not changed much since the 1977 NSCR which indicated that those who own or who have access to a car are almost twice as likely to make trips to the countryside than those who do not. In the 1977 survey, only 14 per cent of trips made in the four weeks prior to the survey were made by those without access to a motor vehicle.

A basic understanding of the social structure of recreation participation is also important as an input to our analysis of recreation social policies in the next part of this chapter. In simple terms, as people's social or occupational status rises, so their propensity to recreate in the countryside increases. This may be seen in figure 6.3. The first two of these figures show that from household surveys, in terms of both occupational status (from the General Household Survey) and social class (from the NSCR), there is a consistent pattern over time, really from 1973 to 1984, of declining participation in countryside recreation as social or occupational group declines.

Importantly for our analysis later, this relationship between social status and level of participation is upheld, from figure 6.4, at a number of managed passive recreation sites. This figure indicates the proportion of people participating at different informal passive recreation sites in each number of occupational groups, relative to the proportion of people in those occupational groups in the United Kingdom as a whole, which is the dotted profile over-laying each of the individual distributions. Again, you will see an over-representation of people in higher occupational groups and an under-representation of people in lower groups indulging in such passive recreation, relative to the national average.

We will now make use of this brief review of recreation behaviour patterns to assess the potentials and limitations of both social and land-use policies for recreation.

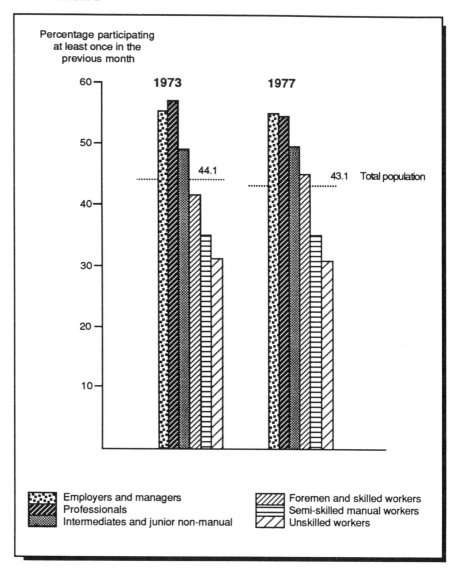

Figure 6.3(a) Social structure of participation. Those from higher status occupational groups tend to be more frequent visitors to the countryside.
Source: General Household Survey and Countryside Commission

Social policies for rural leisure

Ever since the Second World War, there have been government reports, policies and even sections of legislation, that have championed the need to develop social policies for rural leisure – the less well-off should be given the opportunity to recreate in the countryside. We have reviewed a number of the more recent of these in chapter 1. These policies have tended not to work too well. Why is this? We will now examine this issue and consider an example of an attempt at a set of social recreation policies that has been fraught with difficulties.

We can get some idea of the limited effect of social policies from the information on the social structure of recreation participation that we considered above. You will recall that people are much more likely to participate in countryside recreation if they are in a higher status occupation, have good access to a car and are in a higher social group. These observations on social structure indicate that it is demand factors – incomes, education,

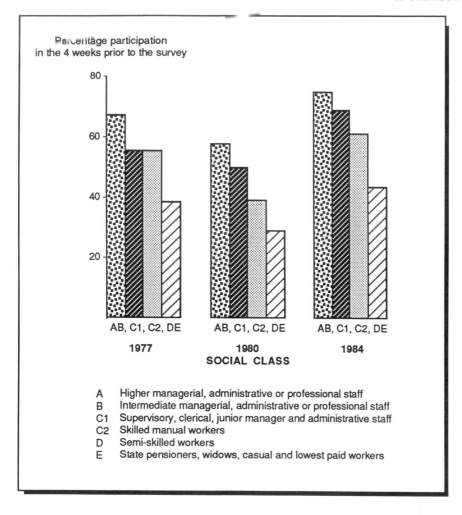

Percentage participation in the 4 weeks prior to the survey

SOCIAL CLASS

1977   1980   1984

A   Higher managerial, administrative or professional staff
B   Intermediate managerial, administrative or professional staff
C1 Supervisory, clerical, junior manager and administrative staff
C2 Skilled manual workers
D   Semi-skilled workers
E   State pensioners, widows, casual and lowest paid workers

Figure 6.3(b)

car ownership and so on – that trigger recreation participation, rather than any expressions of social need that have been met by a public policy response. In other words, there is no evidence from observed participation that social policies have had any significant impact on attracting the less well-off to the countryside.

If there is, then, a lack of progress in developing effective social policies for rural leisure, how could this have been caused? In fact, there are two main reasons that have inhibited the effectiveness of social policies. The first, which we mentioned in the introduction to this chapter, is the preoccupation on the part of all public authorities during the 1960s and 1970s to make sure that the 'recreation explosion' did not get out of hand. The second, which is much more important, is that nobody really stopped to ask if those people who were not participating in rural leisure – chiefly the lower social and occupational groups – did so because they were in some way deprived of the opportunity to do so, or simply because were not very interested in the countryside.

We must, therefore, look briefly at these two causes of the limited success of recreation social policies, before examining the problems of social policies concerned with recreation transport more closely.

Although all public recreation providers were allowed to develop social

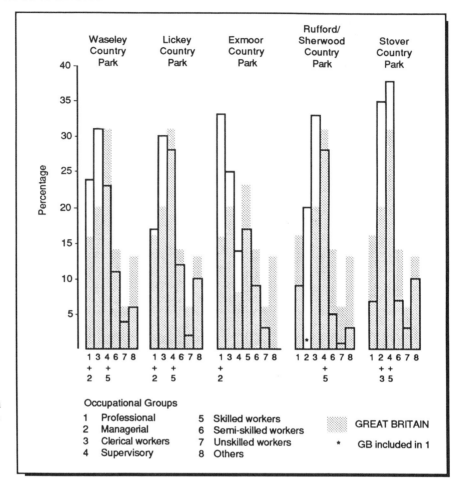

Figure 6.4 Participation structure at specific sites. The occupational structure of participation is supported by evidence from individual recreational sites.
Source: N. Curry and A. Comley

policies for countryside recreation, local authorities and the Countryside Commission in particular were allowed only to facilitate recreation under the 1968 Countryside Act. This meant that they were not able positively to promote countryside recreation. Because of this, they were severely restricted in the amount of 'social targeting' of recreation that they could undertake, since this would be promotional activity. They were only really in a position to cater for existing demands, rather than seek to influence thwarted needs.

This 'facilitation vs promotion' set of responsibilities of providers was really a matter of getting policy wires crossed. The 'facilitation' element of the 1968 Act was designed to limit the recreation explosion, rather than neutralize social policies, but it did have the latter effect too. The impotence of social policies, however, was made even worse by the fact that most local authorities, in abject fear of this explosion, actually sought to control rather than simply facilitate, access to the countryside.

Throughout the 1970s, in their structure plan policies, county councils put forward recreation policies designed to contain people, to keep them out of the deeper countryside, to intercept them and filter them into country parks and generally keep them well clear of all agricultural land. We will look at some of the implications of these policies for rural land uses later in this chapter, but for the present, as one commentator of the time put it, these policies certainly did not sound like they were being designed to help people enjoy themselves. In this context, you can see that the development of policies actually to encourage the less well-off to visit the countryside would have been rather difficult.

It is also hard to develop successful social policies without a clear picture of what social needs actually are. During the 1970s, there were increasingly vociferous views that stressed the importance of developing social leisure, from organs such as the National Council for Sport and Recreation, the Town and Country Planning Association and the English Tourist Board and by those responsible for the Cobham Report on Sport and Leisure and the 1975 White Paper on Sport and Recreation. None of these, however, had done any particularly close research on the nature of leisure needs. They had all assumed, with philanthropic good intent, that the social structure of countryside recreation participation that we have described above showed a lower propensity to recreate amongst the less well-off, because they were in some way deprived of participation.

Clearly, any non-participation in countryside recreation may be due to deprivation, but it may also be due simply to a lack of interest. If there is a significant element of this 'lack of interest' amongst these less well-off non-participants, social policies for recreation designed to counter deprivation, will be misdirected. In fact, there is evidence to suggest that the interest in countryside recreation declines as social and occupational status declines. Let us return for a moment to the first of the National Surveys of Countryside Recreation (NSCR) that we considered above. Coupled with the collection of large amounts of statistical data, this survey also sought views on people's motivations for visiting the countryside. Here, the surveyors found that amongst lower social groups, an interest in the countryside was also lowest.

Many manual workers, for example, found the countryside boring because

Figure 6.5 The country park experience. Channeling people into managed sites, such as the Sherwood Country Park, was a management priority in the 1970s. *Source*: Countryside Commission

there was nothing much to do when you got there. Others, and particularly older people, had unhappy memories of the countryside, often passed down from earlier generations who had left the land to find work in the towns. These recollections were of a countryside where workers received exceedingly low pay for working very long hours and were often dispossessed of their homes when they became unemployed. Such a rural image sapped the interest in the countryside of many lower social and occupational groups. When asked directly about their preferences, there was again a clear correlation between occupational status and the preference to spend leisure time in the town. This is illustrated in figure 6.6 below.

This evidence is supported by other research into the relationship between the social structure of participation and the type of recreation activity provided. Studies in South Wales, Gloucestershire and Yorkshire indicate that the social profile of participation shifts to lower occupational groups when more 'performance' or 'showman' types of activity are put on, and higher occupational groups are more evident only when more passive or solitary activities are available. In these cases clearly, it is preferences rather than constraints that are driving participation (see figure 6.7). Social researchers in London also have related these participation patterns to different attitudes between lower and middle class groups to the education of children – as social status increases so does the felt need to educate children about the countryside. There is also some evidence to suggest that lower occupational groups more generally prefer to spend their leisure time with their peer groups rather than their families, which tends to make countryside recreation a less likely activity for them.

The point of this evidence is, of course, not to suggest that there is no interest in the countryside in certain social groups, nor that certain social groups do not find it hard materially to participate in countryside recreation. What it does show, however, is that policymakers cannot simply assume that all non-participation in countryside recreation is as a result of material deprivation.

So how do these conclusions about the social structure of recreation participation relate to the recent reformation of countryside recreation

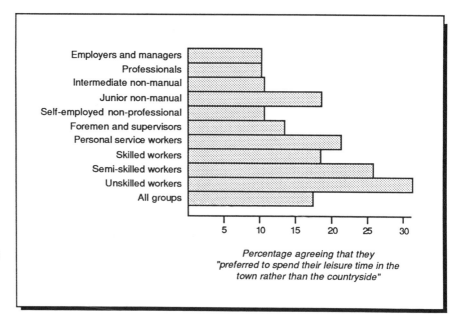

Figure 6.6 Recreation in town and country? Not all social groups have the same inclination to spend time in the countryside.
Source: M. Fitton

policies in the Countryside Commission's 1987 *Policies for Enjoying the Countryside* and other recent policy initiatives outlined in chapter 1? To begin with, most government agencies and particularly the Countryside Commission and the Ministry of Agriculture, Fisheries and Food (MAFF) unlike their duties in the 1970s and early 1980s, now have clear promotional roles in the development of the public enjoyment of the countryside. You will recall from chapter 1, for example, that section 17 of the 1986 Agriculture Act requires agriculture ministers to *promote* the enjoyment of the countryside by the public. Clearly, this promotional stance overcomes what have been slightly ambiguous purposes of public bodies towards countryside recreation and allows the positive development of social policies to the extent that they might be appropriate.

In pursuit of this promotional role, the Countryside Commission's *Policies for Enjoying the Countryside* intends to improve public awareness, public understanding, public confidence and public ability. Wisely, perhaps, they propose to do this without targeting specific social groups and are seeking to develop these skills on the part of the farmer and landowner as well as the recreationist. The means by which they wish to do this, too, represent an improvement on previous policies – something that we shall return to later in this chapter when we consider recreation implementation and management.

There does appear to be a move in current policy, then, away from trying to develop blanket policies for specific social groups towards a more general promotion of the virtues and values of countryside recreation to the public at large. There are, however, certain exceptions to this. The Countryside Commission, for example, has launched a series of experiments around Nottingham under the title 'Operation Gateway', specifically to encourage ethnic minority participation in the countryside. In their *Policies for Enjoying the Countryside* too, they propose the targeting of those without ready access to a car in the future development of public recreation transport policies although, this may meet with less than total success, as we shall see.

Recreation transport policies

We have already gathered from some of the data on countryside recreation participation outlined above, that the motor car is a very important means of getting to the countryside for leisure purposes. This is fine if you own a car, but if you do not have immediate access to one (like some 40 per cent of the population), participation in rural leisure can be quite difficult. It has thus become conventional wisdom that developing alternative means of recreation transport, particularly for the disadvantaged, should be an important policy goal for public recreation. This is principally a policy of social equity. In addition, though, public recreation transport also is felt to ease traffic congestion, push people into the 'real' outdoors more and save energy. It may also help to keep existing rural transport services alive in otherwise marginal situations.

Public recreation transport was thus promoted from the beginning of the 1970s, particularly by the Countryside Commission who set up a series of recreation transport experiments concerned particularly to get people from the larger cities into the open countryside. These types of initiative became adopted by many county and district councils in the structure and local plans of the later 1970s and they were given positive promotion by the Regional Councils for Sport and Recreation in their Regional Recreation Strategies into the early 1980s. In fact, between 1976 and 1983 around half of the county councils in England and Wales directly supported or promoted recreational transport services. Around two thirds of these were new,

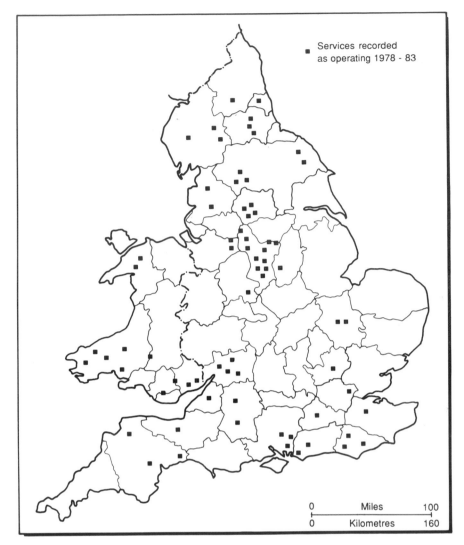

Services recorded
as operating 1978 - 83

0   Miles   100
0   Kilometres   160

*Figure 6.8 Public
transport for rural
enjoyment.* Recreational
transport experiments
have often failed to be
the success anticipated
for them.
Source: D. Groome and
C. Tarrant

specifically recreation services, rather than enhancements to existing routes.
Most of these services, illustrated in figure 6.8, were designed, interestingly
enough, to take people from urban areas to more distant places rather than
the countryside around the towns. Over half of the authorities maintained
that such services were predicated on the notion of social equity, rather than
public preference or any other type of objective.

Of course the problem with the development of such policies, at least in the
pursuit of social equity goals, brings us back to the question of preference or
constraint – were people who did not have access to a car (who have been
considered by policymakers to be the relatively disadvantaged) all wanting to
go to the countryside, or were many of them not really interested even if
transport were available in another form? Recreation transport researchers
at the University of Manchester do suggest that many of the providers of
these alternative transport services were more concerned with where people
might most like to go than whether they wanted to go in the first place.

In a review of such provision by the mid-1980s, these researchers found
flaws in a number of schemes. The failure of many recreation services was not
unusual. Around a sixth of all those started between 1978 and 1983 had been

*Figure 6.7 Time to sit
and think. . . .* Passive
leisure experiences seem
more popular amongst
the better off, according
to national evidence.
Source: Gloucestershire
County Council

abandoned. All of the schemes required subsidies (as you might expect for social policies) and around a half of them failed to do as well as the operators themselves had expected, sometimes carrying very few passengers at all. It was significant, though, that those schemes which were 'added on' to existing rural transport services that served local people or existing tourist areas, did better than those set up specifically for recreation purposes. Significant too, was the invariable absence of marketing studies before the schemes were started. From other evidence too, a Sunday bus service in Gwent was used much less than expected and in Gloucestershire a free bus service from one of the poorer parts of a town to the local country park was withdrawn through lack of support after three Sundays in operation.

A further setback to the success of using public recreation transport as a social equity policy for recreation comes again from the Manchester University study, where they found in individual recreation site studies that it was by no means always lower social groups that were the dominant users of alternative recreation transport, which you might expect if such services were aimed at the relatively disadvantaged. It would seem here, too, that there is an element of 'the disadvantaged' preferring not to visit the countryside rather than simply being deprived of doing so through lack of access to a car.

Clearly, then, there has been a problem in developing recreation transport policies as social equity policies. The Manchester University researchers, though, have made some suggestions about how this policy process might be made more effective. These range from developing community initiatives and promoting cycling and walking, to thinking about locating recreation facilities in more accessible places, rather than trying to transport people to less accessible ones.

During the 1970s, it was thought that urban fringe locations would be much more easy to get to than the deeper countryside, and as a result, policy priorities were given to the location of recreation facilities close to towns. Research since then has shown that there does not appear to be much of a difference in the type of people who visit urban fringe sites compared to more distant ones, and as a result it is now felt by many that the most effective location of 'countryside recreation opportunities' if you wish to target the underprivileged, is the inner city itself. This could be done through the use of urban farms and off-site interpretation centres. None of this will be of much use, though, if people are still not very interested in gaining a countryside recreation experience, so the researchers also suggest that market research is an important element of social policy, if resources are to be used to best effect. This sounds a bit odd – using market tools to develop social policies – but it certainly would assist in distinguishing between those disadvantaged who would like to go to the countryside more, and those who are not very interested. In this context, though, it is important to distinguish between identifying markets to improve provision, and actively trying to sell the countryside to people who are otherwise not interested, particularly if their opportunities are limited.

Despite the limitations in using recreation transport initiatives as social equity policies, the Countryside Commission's 1987 *Policies for Enjoying the Countryside* still seeks to pursue them. They are proposing special services with discounts, timetable adjustments with 'attractive' fares and better publicity and opportunities for such services. History suggests that these new policies might be less than totally successful.

Having traced some of the problems associated with recreation social policies and some of the improvements suggested in current policy initiatives, it is now important to turn our attention to policies for the use of our rural land for leisure.

Land use policies for rural leisure

If social policies for countryside recreation have had their shortcomings, land-use policies for recreation too, have not been without their difficulties. To begin with, there are significant problems in the co-ordination of these policies because there are simply so many governmental agencies with responsibilities for their formulation and implementation. In general terms, as we noted in chapter 1, government responsibility for leisure falls uneasily between no less than six Ministries. Obviously, the Ministry of the Arts and Libraries has a sigificant influence on the use of our leisure time, but the Department of Education and Science also has responsibility for adult education which is chiefly a leisure pursuit. The Department of the Environment (DoE) too, is responsible for the Countryside Commission, the Sports Council and the local authorities, all of whom have leisure responsibilities. The Department of Trade and Industry has responsibility for the tourism sector, and the Department of Employment influences the level of our leisure time, through the development of guidelines for both hours of work and for holidays. Finally, MAFF now has recreation responsibilities on agricultural land under the 1986 Agriculture Act, but it has had duties for much longer than this in our publicly and privately owned forests through the Forestry Commission which itself is part of MAFF.

It is not surprising, perhaps, that with such a broad distribution of leisure responsibilities in government, policies specifically for recreational uses of our rural land also are widely dissipated. Figure 6.10 gives an indication of the principal agencies involved in rural recreation planning. Policy co-ordination, then, is the first problem that we encounter when examining

Figure 6.9 The countryside leisure bus. Despite uncertainties about the extent of their use, such services are still a central part of Countryside Commission policy. *Source:* CoSIRA

recreation land-use policies. Although attempts have been made to make the policies of these agencies more integrated, particularly through the Regional Councils for Sport and Recreation, it is one that is still evident in the late 1980s.

But what of the nature of land-use policies? How do they relate to the outline of people's observed behaviour that we presented earlier in this chapter? To examine this question, we will focus on one particular set of land-use policies formulated by the agencies in figure 6.10 below that has the widest impact in England and Wales as a whole. This is the strategic policies for recreation produced by all county councils in their county structure plans. Recent research for the Countryside Commission, which examined all structure plans and structure plan reviews in England and Wales to the end of 1985 – there were over 100 of them – indicates some interesting characteristics of land-use policies in terms of recreation land quality, and recreation activity types.

In this research, different types of land areas where recreation *was not* to be encouraged in the countryside were classified according to the frequency with which they were mentioned in structure plans. The results are set out in figure 6.11(a). You can see from this that it is high value landscapes and the productive countryside that are the types of area where county councils in general would not like to see recreation taking place. Clearly, there are elements of the fear of the recreation explosion influencing these types of policy again, much as was the case in the development of recreation social policies. There is also an element of agricultural fundamentalism in the thinking behind such policies.

The problems with these two types of priority, however, are threefold. First, not encouraging recreation in areas of landscape conservation value is steering the recreationist away from the most pleasant parts of the countryside, where you might expect his or her satisfactions to be the greatest. Such a policy certainly does not correspond to the sorts of places that the public

Figure 6.10 The leisure policy makers. Policy coordination for rural recreation is complicated by the large variety of agencies involved.
Source: N. Curry

AGENCY	DATE OF FORMATION	ROLE
Forestry Commission	1919	Responsible for recreation in state and private forests
British Waterways Board	1968	Development of amenity and recreational use of British inland waterways
Water Space Amenity Commisssion	1974	Develop the recreation potential of the water industry
English Tourist Board	1969	Encourage tourism and the provision of facilities
Nature Conservancy Council	1972	Recreation in National Nature Reserves if no conflict with conservation
Countryside Commission	1968	Facilitate the enjoyment of the countryside
Sports Council	1972	Sport in the countryside. This includes walking
Regional Councils for Sport and Recreation	1976	Co-ordination of agencies
County Councils	1974	Structure plans, country parks, footpaths, etc
National Park Authorities	1974	Enjoyment of recreation
District Councils	1974	Playing fields, signposts, etc

	ACTIVITIES NOT ENCOURAGED IN GENERAL	SPECIFIC ACTIVITIES NOT ENCOURAGED
AREAS OF LANDSCAPE CONSERVATION VALUE	38	11
Areas of Outstanding Natural Beauty	10	6
Heritage Coasts (and other coastal areas)	10	0
Areas of High Landscape Value	7	1
Unspecified areas of designated landscape	5	1
National Parks	4	3
Sites of archaeological significance	2	0
PRODUCTIVE COUNTRYSIDE	26	4
All agricultural land	20	3
Forestry land	3	1
High grades of agricultural land	3	0
THREATENED AREAS	26	1
Areas of excess or low capacity	10	1
Environmentally sensitive areas	10	0
Vulnerable or fragile areas	5	0
Areas of conflict	1	0
OTHER SENSITIVE AREAS	18	4
Inland water areas	7	2
Deeper or remote countryside	5	1
Open countryside	2	1
Quiet areas	2	0
Attractive or high amenity areas	2	0
AREAS OF NATURE CONSERVATION VALUE	15	3
'Natural' environments (including those of wildlife and ecological significance)	8	2
Statutory nature conservation areas (nature reserves, S.S.S.I.s)	6	1
Conservation priority areas	1	0
AREAS WITH INSUFFICIENT FACILITIES	3	0
Areas with no suitable car parks or with no suitable access	3	0

Figure 6.11a Recreation priorities in structure plans. Keeping visitors away from our best landscapes and agricultural land has been a major concern for local authorities (see fig. 6.11b).

would appear to like to go, as illustrated by figure 6.2(c) of this chapter. Second, there is very little evidence indeed, that recreation pressures in the countryside actually cause much permanent damage at all to areas of high conservation value. Third, as you will be fully aware from chapter 2, not encouraging recreation on farmland flies entirely in the face of current thinking about the need to diversify the agricultural economy and find new uses for agricultural land.

If we look now at the types of land area where receation was to be encouraged by county councils, this picture is reinforced. In figure 6.11(b) it is

	ACTIVITIES ENCOURAGED GENERALLY	SPECIFIC ACTIVITIES ENCOURAGED
ACCESSIBLE LOCATIONS	72	12
The urban fringe	36	11
Accessible sites	13	0
Sites near to main roads	12	1
Sites near to public transport routes	8	0
Sites within walking distance of users	3	0
WATER AREAS	70	16
Rivers and canals	48	5
Inland water areas (not linear)	13	2
Coasts and beaches	9	9
AREAS OF DERELICT LAND	64	5
Disused mineral workings / gravel pits	23	1
Derelict, disused or damaged land	20	0
Disused railway lines	14	4
Low value land	5	0
Areas of land reclamation	2	0
SITES TO ABSORB CAPACITY	42	9
Forest and woodland	12	3
Areas which can absorb capacity	10	2
Recreation priority areas / intensive recreation areas	9	2
Where conflict is minimised or diversion takes place	6	2
Historic houses	2	0
Lowland areas	1	0
Linear open space	1	0
Buffer zones	1	0
AREAS OF HIGH NEED OR POOR PROVISION	14	0
HIGH VALUE LANDSCAPE	9	13
Green Belts / Green Wedges	6	0
Attractive areas	2	0
Deeper or open countryside	1	4
Areas of Outstanding Natural Beauty	0	4
Upland areas	0	2
National Parks	0	2
Heritage Coasts	0	1

Figure 6.11b As with fig. 6.11a the numbers in the tables indicate the frequency with which each area is mentioned in all structure plans and their revisions up to 1985. *Source*: N. Curry and A. Comely

very much ordinary and low value landscapes that are to be the rural recreation areas of the population, and high value landscapes are again to be avoided. Encouraging recreation in areas of derelict land may be particularly unfortunate since, although they may be undistinguished in landscape terms, in many cases they are quickly becoming some of the richest wildlife habitats in the country, precisely because of their dereliction.

Figure 6.12 A walk on the wild side? Derelict sites can be ecologically rich and therefore not as durable as places for a range of leisure activities as local authorities think them to be.
Source: T. Edgar

In terms of land-use *quality* for recreation, then, certainly as far as structure plans are concerned, there is evidence to suggest that priorities for recreation may not entirely correspond to either public preferences or public policies in other sectors of the rural economy. But are policies for recreation *activities or facilities* on this land any more sound?

Finally from this research into structure plans, figure 6.13, indicates the degree of encouragement that is to be given to different types of recreation facility or activity in the countryside. Activities are either to be permitted, promoted or provided by public authorities. You will see from this that it is generally the more passive types of recreation that are to be encouraged by county councils rather than the more active. Again, this does not correspond particularly closely to our understanding of people's behaviour. Country parks and picnic sites, for example, which are amongst the most popular forms of provision by public authorities, are only a minority destination for the population at large as we can see from figure 6.2(c). Again from what we said above on the social structure of participation, there is evidence to suggest that it is the more passive forms of recreation that tend to be of more interest to the higher occupational and social groups, rather than the less well off. More importantly, though, the more active or sport-orientated an activity becomes, the less likely it is to be encouraged by a county council. This is precisely the opposite of the relative growth in popularity of different types of recreation activity that we outlined earlier in this chapter.

So there are also some shortcomings in recreation facility and activity policies for our rural land. This notion is given further weight by the fact that this same research found that it was the availability of grant-aid for certain types of recreation project, that determined the extent to which they would be provided, rather than any notions of public preference.

Nationally too, it has been the broad policy of public agencies to concentrate on the provision of specific facilities. However, the evidence on types of recreation activity that we outlined earlier in this chapter has caused a significant review of this position. In recognition of the fact that only a small proportion of visits are to fully-managed areas such as country parks and

	PERMIT	PROMOTE	PROVIDE
INFRASTRUCTURE	0	1	37
Information and interpretation	0	0	22
Car parks, toilets, unspecified	0	1	15
INFORMAL PASSIVE RECREATION	28	45	99
Country parks	6	13	37
Picnic sites	1	6	37
Scenic drives or routes	0	2	9
Visiting historic assets	1	9	6
Informal recreation (unspecified)	5	2	6
Moorings and marinas	9	7	0
Cruising	4	3	2
Visiting places of interest	0	1	0
Birdwatching	2	2	2
INFORMAL ACTIVE RECREATION	6	15	74
Short walks / footpaths / rights of way / access	3	12	61
Small scale facilities (unspecified)	3	3	13
FORMAL ACTIVE RECREATION	24	17	29
Horse riding / pony trekking / bridleways	5	3	11
Active outdoor facilities (unspecified)	10	1	2
Unspecified water based	6	7	7
Hill walking / fell walking / rock climbing / Long Distance Routes	3	6	9
SPORTS - LOW CAPITAL	25	13	13
Golf	20	4	0
Fishing	3	5	4
Field sports	2	1	0
Orienteering	0	1	0
Cycling	0	2	9
SPORTS - HIGH CAPITAL	15	14	6
Sailing	3	3	2
Rowing / canoeing	1	4	3
Gliding / hang gliding	5	2	0
Specialist sports (unspecified)	2	2	0
Water skiing	1	1	0
Ski slopes / skiing	1	0	1
Motor racing	2	0	0
Power boats	0	1	0

Figure 6.13 Recreational facilities in Structure Plans. Passive rather than active recreation is encouraged by local authorities. Numbers are, as in fig. 6.11, representations in Structure Plans.
Source: N. Curry and A. Comely

picnic sites, the Commission's 1987 *Policies for Enjoying the Countryside* now seeks a much broader focus on recreational land areas, to include specific attention to, amongst others, Long Distance Routes, access to land and water, managed recreation sites, village and rural communities and landscape conservation areas.

The Commission's land area proposals include the maintenance of some specifically managed sites as gateways to the deeper countryside, but also embrace the more active development of opportunities presented by less formal 'access areas' and routeway systems which may be linked together and used for cycling and horseriding as well as walking. Consistent with the areas in which recreation is to be encouraged in county structure plans, illustrated in figure 6.11(b) above, the Commission also is proposing the development of new forest areas close to towns, to be both accessible locations and sites which will absorb capacity. This is consistent with some of the proposals we discussed in chapter 3.

Recreation also is to be encouraged in villages as long as it assists in the development of the village economy. In designated conservation areas or areas of high landscape value too, recreation is now to be encouraged rather than deterred. Aggressive management practices are now to be used as the means by which these precious landscapes will be conserved: we review these briefly below. In respect of both land use and social policies in structure plans, the Commission calls for a reappraisal. Recreation policies in structure plans should be more fully integrated with policies for the countryside. The Countryside Policy Review Panel of the Commission would even like to see the development of co-ordinated regional plans for recreation, produced by local authorities jointly.

The new policy proposals generally, then, move some way towards overcoming some of the main problems associated with the development of recreation land-use policies. In one salient respect these national policy guidelines differ from all of the national and structure plan policies of the 1970s and early 1980s. They seek to take farming and landowning interests in partnership in the development of public enjoyment over productive rural land and no longer seek to treat agricultural land as sacrosanct.

Despite these new proposals for the development of countryside recreation, two other aspects of recreation land-use policy still remain problematic – those of rights of way and common land. Rights of way are probably the single most important recreation resource in the countryside. We have about 120,000 miles of them in England and Wales, but only a small proportion of

Figure 6.14 Signposting rights of way. Venturing off the beaten track is for many dependent on clearly marked paths. *Source*: N. Curry

Figure 6.15 'He's right you know luv – there is a public right of way through here.'
Source: *Western Mail*

these actually are in use. Many paths have been obstructed or ploughed up to the extent that the public is very cautious and uncertain about their use. Many will use only footpaths that are clearly signposted (see figure 6.14).

Rights of way also enjoy a peculiar administrative system. Whereas most countryside recreation responsibilities fall to county planning departments, rights of way traditionally have been the jurisdiction of county highways authorities and there is often less than perfect communication between the two. Many local authorities are not very assertive about enforcing rights of way, and a number do not even keep an up-to-date record of them. (This record is known as a definitive map.) Rights of way too, are often the most controversial of all recreation facilities – recreation conflicts are often characterized by the farmer chasing the recreationist off his land where the public right of way is in dispute. Farmers have also been historically reluctant to allow changes to public rights of way for fear of greater public access to their land.

All of this leads to a great degree of under-use of public rights of way, and because of this, the Commission's *Policies for Enjoying the Countryside* is making the legal definition of rights of way and their up-to-date monitoring through definitive maps, their top priority. This will involve the establishment of liaison groups between the public, local authorities and farmers and a significant input of both resources and manpower. The rights of way network eventually should be fully co-ordinated.

Common land

Common land is land over which the public have certain long-standing legal rights, usually associated with access (but sometimes with grazing and mining). This land ranges from vast upland areas to small urban pockets and village greens. At present, however, only 20 per cent of the 625,000 million hectares of common land is legally open to the public and few commons have any form of active management on them. They are thus areas of both recreation conflict and environmental neglect. Despite these problems, many attempts to clarify the legal position of common land have failed. A variety of Private Members Bills have come before Parliament (in fact as we have noted in chapter 1 there was one in each of the first six years of the Thatcher administrations), but none of them have ever successfully proceeded to an Act.

The Countryside Commission has been active in this area of recreation land-use policy too. In 1984, it set up the Common Land Forum representing all interests in common land, to work out suitable management, protection and legislation measures for commons. It, too, points to the need for new legislation. The main conclusions of the Common Land Forum, reporting in 1987, concern the setting up of management associations between owners, local authorities and the public, to work out suitable five year management plans. After this period, all commons should be open for quiet public access, but this should be subject to regulation in the interests of conservation and farming. Such regulation would pertain to such things as horse riding and to dogs. Village greens should be fully available for local sports and pastimes (see figure 6.16) and all commons should be appropriately registered. These

Figure 6.16 The village common – Exford Green, Exmoor. Unrestricted access to village greens is called for by the Common Land Forum. *Source*: Countryside Commission

proposals represent an unusually comprehensive degree of consensus amongst all of those with an interest in common land, but it still remains to be seen whether these recommendations will find their way into the statutory framework widely considered necessary to make them effective.

So far, we have examined both recreation social and land-use policies in some detail. Let us now turn to consider more briefly some particular aspects of how recreation policies are implemented and how recreation sites are managed.

Management policies for rural leisure

The translation of national and strategic policies for rural leisure into an implementable framework falls to the recreation manager on the ground – the ranger, the warden or project officer. These are usually employed by local authorities or the voluntary sector and in line with the cautious policies for recreation during the 1970s and early 1980s, many of them have a predominant concern for environmental conservation and the interpretation of the conserved environment. We will examine below some of the problems associated with management policies for information and interpretation, but let us look first at the more general current thinking about recreation management and implementation. Again, from the Countryside Commission's *Policies for Enjoying the Countryside*, the predominance of conservation concerns has been replaced by a threefold set of objectives – the management of the land; the management of the visitor; and the management of public rights of way.

The management of the land is still concerned with the conservation of the physical environment, but also with the minimization of the conflict between the landowner and the visitor. Such a role is to remain shared by the local authority and voluntary sectors. Managing the visitor, however, is now seen as the primary role of the ranger or warden and this will require the development of new interpersonal skills and new functions or liaison between interest groups and the local community. A major training effort for these types of staff is associated with the new range of implementation policies. The Commission's proposals for implementation concerning rights of way have been considered above.

But as an illustration of how implementation and management policies can be misdirected, and how current policies are helping to put things right, let us examine one particular case of policies for implementation and interpretation.

The Countryside Commission, local authorities and a number of other organizations have always accorded information and interpretation a high priority in recreation policy. This has been driven by notions of wishing to give the public something that will do them good, and to the extent that this is a philanthropic ideal, such policies could be considered to have a social policy component to them. In addition, such policies have been driven by the managers of countryside recreation sites who, as we noted above, were often trained in subjects that made interpretation a high priority. During the 1970s the Commission published a number of studies into both the benefits of interpretation and into the development of interpretation packages themselves. Unfortunately, one area that was not closely studied, as in the case of the transport policies we discussed above and the pricing policies that we shall consider below, was what people actually wanted from interpretation and information services. Without such an understanding, such philanthropy was always in danger of being misplaced.

Some research has since been undertaken into the popularity and usefulness of these services, and although it is undoubtedly partial because it is based on site evidence only, it does suggest that on-site interpretation facilities have not been too popular (see figure 6.17). In a survey of Lochore Meadows Country Park, for example, it was found that although 60 per cent of participants did recall interpretation leaflets being at the Park, only 30 per cent claimed to have read them. The usefulness of interpretation facilities has also been studied at Crickley Hill Country Park in Gloucestershire. Around 40 per cent of those surveyed considered sign board information to be useful, but there were very high degrees of non-use (around 80 per cent) of the visitor centre, leaflet information and wardening services. This lack of use was invariably reinforced by the reasons that people expressed for visiting the park, which, more often than not, did not require the use of any ancillary information.

Specifically for countryside visitor centres (where you might expect interpretation facilities to be most well received), further research has shown that certain sectors of the population appreciate such services more than others. Intermediate managerial occupations and members of organized clubs seem to get the most out of interpretation facilities, but these are also found in the research to be those that are already the most well informed about the countryside. Despite this difference in people's interest, this visitor centre's study concluded that information and interpretation appear to have very little long-term effect on the visitor.

Finally, psychological studies in the North Yorkshire Moors National Park have found that on-site interpretation attracts a very narrow social sector of the population (which, in fact, mirrors that of the interpretation provider). It also appears to have an over-emphasis on presentation rather than communication, and hardly serves to educate the visitor at all. In this particular research study, the author concluded that the current approach to on-site interpretation actually serves to alienate lower social groups who undertake recreation participation.

Figure 6.17
Interpretation on the North York Moors. Interpretation displays like this may look good, but can appeal to a small minority.
Source: Countryside Commission

The Countryside Commission's 1987 *Policies for Enjoying the Country-side* pursues the theme of information and interpretation facilities in the context of the need to improve public awareness of the countryside. A study conducted by them shows that there is a considerable demand for more information about what to do and where to go in the countryside once you get there. The Commission is keen to enhance such general information availability but it is now promulgating what is a central characteristic in this policy area. Importantly, information is now seen as being required about the countryside, not actually in it, but in the places where people live and work. Such information dissemination should be done in partnership with news-papers, local radio and other established information services, rather than separate from them. This is important when considered in the context of other research findings. The recreation transport research discussed above concluded that social equity policies might be best served, in transport terms at least, by taking the 'countryside' into the towns rather than trying to get people away from their local surroundings. The visitor centre research, too, indicated that the success of interpretation facilities depended on their location and this location need not be anywhere near what was being interpreted.

The Gloucestershire research, too, concludes that interpretation fails when it does, not because it is doing the wrong things, but because it is in the wrong place. When people go to the countryside for an outdoor experience, reading leaflets, sign boards etc. may well not be part of that experience. Interpretation of the countryside in schools, evening classes and other urban centres, however, may be a desired substitute for or complement to that 'open air' pursuit.

Policies for information and interpretation then, and particularly their location, provide good examples of where finding out what the public wants through surveys and research has helped to reorientate policy successfully. The Countryside Commission now even proposes the development of countryside centres in the larger towns to embrace an information service, media contacts, books and so on. Since the 1968 Countryside Act charged the Commission with caring for people's enjoyment of the countryside such enjoyment has been considered possible only in the countryside. A full understanding of people's preferences has shown that an enjoyment of the countryside can take place quite happily in the towns as well.

From this assessment of the potentials of information and interpretation policies, we now turn to the critical issues of the economic potential of leisure for rural areas.

The income and economic benefits of rural leisure

It is now widely accepted that it is often quite difficult to generate large sums of income from countryside recreation, although it is somewhat easier with rural tourism, a point we discussed in chapter 5. Part of the reason for this is that it was made illegal under the 1968 Countryside Act to charge for access to local authority recreation sites. This was intended as a social equity policy, known as the 'free access criterion' to allow the less well off not to be deterred from countryside recreation by having to pay an entry fee. Clearly this has served to limit the income generated from local authority sites, but it also provides another example of how social equity policies can be misguided, and in the end, can do more harm than good.

The free access criterion was intended to allow the less well-off un-restricted access to public recreation sites. No doubt this was influenced by the types of access that have evolved for urban municipal parks, but it did lead to certain inconsistencies, since other arms of government, for example the DoE, had always charged for access to their countryside sites – in this case designated ancient monuments. Not only were some publicly-owned countryside sites charging a fee and others not, but the less well off were not benefiting from free access in the way that the Act had intended. The social structure of participation that we discussed above indicated that it is the better off that participate in countryside recreation (including visits to local authority recreation sites) proportionately more than the less well-off. Thus it is the better-off that benefit more than proportionately than the less well-off from the free access criterion. A policy of good intent has actually become regressive as well as restricting the income potential of such sites.

Aside from the preference to (or not to) recreate at these sites that we discussed above, part of the reason for the failure of this free access policy is that unlike urban parks, any entry fee to a site is only a small proportion of the total cost of participating in recreation. Most participants will have at the very least transport costs associated with a visit and in the light of this, an entry fee is perhaps not a critical factor in determining participation. This line of argument is supported by a study that was carried out by researchers at Aston University into pricing at National Trust and DoE properties. They found that, within certain limits, visitors to such sites were not too bothered about how much they paid to get in, certainly not to the extent that it put them off going. In technical terms, recreation site entry fees are relatively price inelastic.

The non-critical nature of entry fees also is suggested from some evidence presented above. We noted there that lower social groups tended to visit the countryside more often when 'performance' or 'showman' types of activity were put on, but higher social groups predominated when recreation activities were more passive. These types of activity also correlate with the incidence of charging at sites. It is the 'performance' or 'showman' activities that invariably command a fee and the more passive activities that do not. From this, it could be implied that the less well-off or the lower social groups are not in the least inhibited by entry fees per se, since they are quite prepared to pay them for activities for which they have a preference.

So the free access criterion is socially regressive, and if it were abandoned, the charging of reasonable prices would probably not deter people too much from participating. But does this mean that we can generate significant incomes from rural leisure simply by charging at recreation sites? This is less than certain because there is yet another reason why it is difficult to generate significant incomes from rural recreation. This derives from the fact that such recreation displays elements of what the economist calls a 'public good'. In essence, this means that it has attributes (known technically as 'externalities') that cannot be properly charged for one reason or another. The most important of these in the case of recreation is what is known as non-excludibility. As the name suggests this is where it is very difficult to 'catch' the recreationist to ask him or her for some sort of fee. It would be impossible, for example, to charge people for access to a National Park, because there are simply so many points of entry by road and on foot that you would not be able to channel people through specific access points. A second type of externality is that many people have a legal right of free access that precludes charging. You will be aware of this from the earlier sections on rights of way and common land. In both of these cases the general public has a legal right to recreate free and thus attempting to procure payments would be unlawful.

In the light of these 'public good' characteristics of recreation, how can income generation from rural leisure be developed? There are two principal means of doing this. The first is through some form of public subsidy, and the second is by trying to internalize these 'externalities' as far as possible, to make recreation more of a market commodity. This inevitably means developing tourist enterprises of some sort. We have had some form of policy towards public subsidy for recreation provision ever since 1949 when access agreements were introduced. These allow farmers and landowners to receive cash payments in exchange for an agreement of access over certain parts of their land. There is usually some agreement over the management of that area too. These have not been used too widely outside the Peak District National Park, but other more recent forms of agreement and custodianship payments may make such a means of income generation in a context of agricultural policy change and land set-aside, more common.

We will be looking at the developments associated with management agreements in some detail in the next chapter, but as constituted under the 1981 Wildlife and Countryside Act, these may have an income potential for farmers, when associated with some form of leisure activity. Tax exemptions, too, can be associated with recreation provision in certain areas. The 1976 Finance Act allows for Capital Transfer Tax exemptions on what are called 'Heritage Landscapes', which are identified according to their landscape value. The tax exemptions are invariably tied to some formal access plan for the exempted land. Income generation through some form of public subsidy, then, is likely to develop in the future as a means of increasing the recreation potential of rural land.

As an alternative, how can we set about reducing the externalities associated with recreation so that charging becomes a realistic possibility? The first means, as an attempt at overcoming the non-excludability external-ity, would be, to put it rather crudely, to fence off the recreation area where this is possible. This could only really work successfully for smaller sites, but even where it can be done, problems with charging may remain. A principal problem here, is that revenues simply may not be sufficient to cover costs. Research in Gloucestershire has shown that for smaller country parks, the costs of collecting entry fees are often more than the revenues collected. This is largely due to the fact that the majority of visitors to such sites come at the weekend, when the wages of site staff are at their highest. In these cases, the use of a donations box may provide an alternative means of raising revenue, but these commonly suffer from problems of vandalism.

If charging at sites through the ability to exclude people does become feasible, then income generation will be beneficial to the rural economy, but in these instances there are other advantages of pricing too. It can be used as a rationing mechanism to regulate use and thus protect sites from too much visitor pressure. It can also provide information on levels of use (if you keep some accounts) and thus help to identify what people value.

Other ways of making recreation more of a 'market' commodity relate to its development in connection with some kind of tourist enterprise. As we suggested in chapter 5 many farmers are now introducing bed and breakfast facilities on their farms, and these are often associated with some form of farm holiday where the recreationist becomes actively involved in the work of the farm. Here, access to farm lands which might otherwise have been free is internalized into the cost of a holiday (see figure 6.19).

The Countryside Commission's *Policies for Enjoying the Countryside* document also develops suggestions for the raising of revenues from countryside recreation. The village is one recreation destination for recreationists that might be commercially exploited. This is most often undertaken through retail outlets and through the controlling of, and charging for, parking. In some areas of intense pressure though, such as the

Figure 6.18 A Heritage Landscape – the Pembrokeshire Coast. Capital transfer tax can be reduced for private owners of landscapes like these in exchange for access agreements. *Source*: Countryside Commission

Figure 6.19 Farmers will have to innovate to realize significant incomes from countryside recreation. Source: The Field

Bourton-on-the-Water and Stow-on-the-Wold in Gloucestershire, the feasibility of actually charging for access to the village itself has been considered.

The same policy document also has made proposals for the development of very simple accommodation, usually associated with some form of recreation route, to supplement farm incomes and to utilize redundant farm buildings. It proposes the development of a network of both camping barns – weatherproof shelters with basic amenities –and what it terms bunkhouse barns – a slightly higher grade of accommodation – throughout the countryside.

Clearly, then, there are problems and limitations in the development of rural leisure as a means of providing adequate incomes in a changing rural economy. Public policy however, is responding to both the need for public subsidies for recreation provision, and to developing new markets for recreation, driven particularly by the need to diversify the rural economy away from agricultural production.

Rural leisure to the turn of the century

It is now widely accepted that the recreation 'explosion' of public incursions into the countryside did not actually come to pass in the way that many people had feared. There are certainly fluctuations in the levels of countryside recreation from one year to the next, but this is influenced more by the weather than any significant socio-economic characteristics. In terms of longer-term trends, then, overall levels of recreation activity appear to be fairly constant and are likely to remain so into the foreseeable future.

Despite a reasonably static level of participation in rural recreation, the overall importance of leisure in the rural economy is likely to increase. This will be due principally to the need to find new uses for agricultural land, and new incomes for farmers. Public policies in respect of the economy are likely to include both an increase in the public subsidization of rural leisure, along the lines that we have discussed above, and an increased interest in an effective means of developing countryside recreation as a more marketable product and a revenue earner in its own right. Because of the importance of the changes taking place in other sectors of the rural economy, however, it would seem unlikely that rural leisure will be given the importance accorded to agriculture or forestry, as the Countryside Policy Review Panel would wish.

To an extent, problems may well persist in the development of social policies for rural leisure. Although the Countryside Commission's 1987 *Policies for Enjoying the Countryside* resists targeting specific social groups in all areas but recreation transport, in the development of promotional policies other organizations such as the Chairman's Policy Group that we mentioned in chapter 1 and the Association of Metropolitan Authorities continue to champion the causes of particular disadvantaged groups. Clearly this is not in itself wrong, but there will always remain the danger of a policy mismatch if only poverty and not preferences are taken into account.

In terms of land-use policies, the orchestration of the rights of way issue may well become very important amongst both central government and local authorities alike. It is likely to create a significant number of new jobs and, if successful, resolve some of the principal conflicts currently surrounding rural leisure. It seems likely too, that some form of legislation concerning common land will find its way through Parliament in the not too distant future.

Policies for land uses at a local authority level seem likely to become more consistent with both current policy thinking nationally and with observed recreation behaviour as structure plans, local plans and national park plans are all revised and up-dated. The Countryside Policy Review Panel's proposals for fully-integrated regional recreation plans may well develop too.

Two points of caution should be expressed in connection with the development of new policies for rural leisure. The first of these is that care should be taken in the co-ordination of policy initiatives across the wide range of public agencies that have a responsibility for rural recreation. Historically the large number of these agencies, as we have indicated above in figure 6.10, has led to inefficiencies in policy development and conflict and confusion in policy implementation. It was even the case in 1987 that two policy documents – *Policies for Enjoying the Countryside* and the report of the Countryside Policy Review Panel – both announced new recreation policies separately, despite being set up by the same Government quango, the Countryside Commission.

Finally, what is most important in the development of new policies is that the preferences of the public are taken into account fully. In the past, there has been a paucity of information about public recreation behaviour and this has led to policy formulation that in a number of instances has been counter-productive. It is important, too, that public participation in policy formulation is not mistaken for the expression of public preferences, since this participation often involves only active interest groups who often express only a minority view.

Basing the development of policies for rural leisure on a solid foundation of empirical information about people's expressed wishes may not lead to the provision of recreation for all, but will allow provision for all who want it.

Figure 6.20 But is the enjoyment of the countryside really a marketable *commodity?*
Source: *Farming News*

Chapter 7

Conservation – more than appeasement?

Conservation and politics

As recreation policies are beginning to develop on the basis of observed behaviour and the preferences of our population as a whole, national policies for countryside conservation continue to be determined more directly in connection with the political process. This is partly because such policies are inextricably linked with changes in agriculture and these changes themselves, as we have seen in chapter 2, are frequently determined by political horse-trading in Europe. We have presented a flavour of the political and parliamentary 'debates' that were taking place in the first two terms of the Thatcher government, in the Preface of this volume. We have also looked briefly at some of the countryside conservation impacts of agricultural policy changes in chapter 2. We now develop the analysis of conservation policies in Britain from these two starting points by reviewing the processes involved in the implementation and revision of the 1981 Wildlife and Countryside Act,

From these we go on to examine how the changes in agricultural policies analysed in chapter 2 have led to political victories for the conservation movement and a range of new measures to develop the role of conservation in agriculture. We also assess how agricultural interests have been tempering their activities and developing conservation initiatives. From there we review the conservation consequences of the new initiatives in forestry that we have evaluated in chapter 3, before concluding on the way in which both agriculture and forestry are likely to develop in conservation terms, to the turn of the century.

Revising of the wildlife and countryside act

Environmental groups were undoubtedly encouraged by the increased importance that seemed to be accorded to countryside conservation, certainly from 1984 onwards. This led them to begin, in many cases, to look beyond specific conservation measures to the reform of agricultural policy. But despite these longer term strategic goals of the conservation movement, there was still the immediate task of monitoring threats to wildlife habitats and landscapes and of publicizing the shortcomings, as they saw them, of the 1981 Wildlife and Countryside Act.

We examined the passage of the 1981 Wildlife and Countryside Act in some detail in *The Changing Countryside*. This process was one at which the politics of conservation and agriculture had really come to the fore. The Act itself had had a stormy eleven-month passage through Parliament. Though the conservation lobby had achieved significant concessions, there was considerable disappointment with the legislation which many considered did not measure up to the scale of the problems facing the countryside, particularly from agricultural change. Even at its passing, pressures were mounting to have it revised.

The provisions of the Act included enhanced protection for a wide range of animal, plant and bird species, but it was the provisions for countryside conservation that were most controversial. The Act's central principle was that conflicts between agriculture and landscape or habitat protection should be resolved by voluntary means, backed up where necessary by management agreements. The Act enlarged on and standardised earlier legislation empowering conservation authorities to enter into and finance management agreements with farmers and landowners.

A breach of the voluntary principle was the requirement, added to the Bill in Parliament after intense lobbying by conservationists and over the opposition of the farming lobby, that owners or occupiers of Sites of Special Scientific Interest (SSSIs) should give the Nature Conservancy Council (NCC) notice of their intention to carry out any potentially damaging operations. This, it was hoped, would give the NCC the opportunity to persuade farmers to modify their plans or to negotiate a management agreement. However, considerable contention surrounded one of the Act's provisions for payments to farmers: that where a farmer was refused a capital grant by the Ministry of Agriculture, Fisheries and Food (MAFF) on conservation grounds, the objecting authority (the NCC in SSSIs and county planning authorities in National Parks) was required to compensate the farmer.

Not surprisingly, therefore, the greatest controversy initially surrounding the Act focused on the amount of money that was needed or that had been made available for its implementation and the related issue of how the conservation agencies, particularly the NCC, would discharge their new powers and duties. Environmental groups, however, feared that, lacking the resources to finance more than a few management agreements, the conservation agencies would be reluctant to press their objections to harmful agricultural and forestry schemes, which would therefore proceed unchecked.

A related concern was that the NCC's wariness of antagonizing local farmers was leading to considerable delay and even reluctance in the designation of new SSSIs particularly where forestry or agricultural development was in prospect. First Friends of the Earth and then the Royal Society for the Protection of Birds threatened the NCC with legal action if it failed to fulfil its statutory duty of designating land which met its scientific criteria. These pressures, as well as the unstinting efforts of conservation groups to publicize each and every threat to wildlife sites of any importance, helped harden the NCC in its resolve to use the powers of the Act to the full to safeguard SSSIs. In consequence, the amount of money allocated to site acquisition and safeguards rose sharply. The NCC has estimated that eventually, with 6,000 sites covering just over 8 per cent of Britain and a third of them needing management agreements, the ongoing cost to it of maintaining the system will be between £15 and £20 million per annum.

The other specific concern that environmental groups had regarding the 1981 Wildlife and Countryside Act was the weakness and unwieldiness of its statutory safeguards for protecting threatened sites of conservation value. It remained a bone of contention that these only applied to wildlife sites (i.e. SSSIs) and not to valuable landscapes. But even within this restricted compass, conservationists regarded the Act's formal powers and procedures as woefully flawed and cumbersome.

For a start, the safeguards – including the requirement for farmers to notify the NCC of their intention to carry out a potentially damaging operation – did not begin to apply until a site had been re-notified or freshly designated under the Act. For the existing 4,000 SSSIs, therefore, the NCC faced the task of re-notifying their 30,000 or so owners and occupiers. Each site's wildlife value had to be re-appraised, and then a detailed site description and map as well as a site-specific list of potentially damaging operations was sent to each owner or occupier, and to local planning authorities, water authorities and the Secretary of State for the Environment. Finally, the site had to be registered as a land charge. A major proportion of the time of the NCC's regional staff during the first six years of the Act's operation was taken up with renotification. Indeed, the task will not be completed until 1989.

Inevitably, the NCC's other responsibilities have suffered, including the designation of new sites. These new designations now had to undergo a more complicated procedure before they could enjoy any protection, including a three-months' consultation period to allow owners and occupiers to register any objections before formal confirmation of the notification. In the meantime, the site enjoyed no safeguards whatsoever. Indeed, with the consultation period inviting pre-emptive destructive action on the part of farmers and landowners, it is no wonder that it became known as the three months' loophole.

In 1984, the first systematic statistics became available indicating the effectiveness of the Act's site safeguards. Information from the NCC's regional offices revealed that during the year to March 1984 damage was known to have occurred on 156 existing and proposed SSSIs. This was 3.7 per cent of the total number of sites – undoubtedly an underestimate of the true extent of damage, because with the majority of sites awaiting re-notification, information was patchy. Damage varied from the very minor to total loss (three sites). In over half the cases it was caused by agricultural activities such as ploughing, drainage, re-seeding and the use of fertilizers and chemical sprays. As evidence accumulated of the costs, difficulties and complexities of the Act, pressure began to build up to remedy its most obvious defects. On 30 October 1983, the Council for the Protection of Rural England and the Council for National Parks marked the Act's second

Figure 7.1 Sites of Special Scientific Interest. Designated areas, like Strumpshaw by the River Yare, were at the centre of the debate to revise the 1981 Wildlife and Countryside Act. *Source*: Countryside Commission

anniversary by issuing a statement declaring that experience had proved it to be 'toothless'. Indeed, far from protecting the rural environment, the Act was charged with 'stimulating new pressures on the countryside because of the rights it gives to farmers of large compensation payments'. In the months ahead other groups followed suit with critical reports on the implementation of the Act.

On 15 March 1984, the new chairman of the NCC, William Wilkinson, wrote to the junior Minister, William Waldegrave, requesting that steps be taken to close the three months' loophole and to overcome two other difficulties. The Act required owners or occupiers of SSSIs to give three-months' notice of any intention to carry out potentially damaging operations. In a number of cases the NCC had found this gave it insufficient time to negotiate with the farmer appropriate modifications or a management agreement, and it sought to extend the period of notice. In addition, the NCC was concerned that it was taking up to three weeks to prepare, and for the Department of the Environment (DoE) to process, a nature conservation order – the longstop measure whereby a farmer could be prohibited from destroying an SSSI. A number of sites had been damaged because of this delay in providing the urgent protection supposedly offered by the Act. The NCC, therefore, proposed that it should have an emergency 28-day 'stop' power to cover the period in which the Secretary of State reached his decision on a nature conservation order.

It was no longer possible, in the light of these glaring deficiencies, for ministers to maintain the line that the Act must be given time to prove itself. The Treasury was also beginning to express unease at the escalating cost of management agreements. In March 1984, in the context of the Halvergate controversy, William Waldegrave announced that he was asking MAFF to come up with ways of protecting valuable sites without invoking those sections of the Act which obliged conservation agencies to make substantial payments to landowners.

Ministers and their advisers now directed their efforts towards devising ways of patching up the Act that would appease its conservationist critics and curb the runaway costs. Though now committed to closing the glaring loopholes in the Act, the government continued to drag its feet on this matter, pleading lack of parliamentary time. It seemed also that there was some reluctance to introduce a modest amending Bill for fear of giving MPs and

Figure 7.2 'It's from the Nature Conservancy Council. The yard's been declared a Site of Special Scientific Interest.'
Source: *Farmers' Weekly*

Figure 7.3 'Its to keep farmers off.'
Source: *Chartered Surveyors Weekly*

conservation groups an opportunity to press for more wide-ranging reforms. In the event, a Wildlife and Countryside Amendment Bill was introduced by Dr David Clark, the Labour spokesman for the environment. Dr Clark had consulted widely, and the measure enjoyed the support of the National Farmers' Union, the Country Landowners' Association, the Ramblers' Association, the Council for the Protection of Rural England, the Council for National Parks, Friends of the Earth, the Royal Society for the Protection of Birds and the Royal Society for Nature Conservation. The Bill, which had its second reading on 8 February 1985, would close the three months' loophole and would extend from three months to four the period of notification for potentially damaging operations in SSSIs. A few other specific amendments were also included, the most far-reaching (drafted by the Council for the Protection of Rural England and the Council for National Parks) being to place an additional duty on agriculture ministers and the Forestry Commission to further conservation. The government, while anxious to see the three months' loophole closed, did not support the latter measure and, therefore, there was some uncertainty whether sufficient parliamentary time would be made available to ensure the Bill's passage. However, to sink it would undoubtedly cause the government much political embarrassment. Instead, they ensured that it was emasculated in Committee.

The government's position, that minor adjustments to the legislation were all that was needed, was, in any case being overtaken by other events. A few days before Dr Clark introduced his Bill, the House of Commons Select Committee on the Environment reported on its investigations of the operation and effectiveness of the Wildlife and Countryside Act. Throughout the Autumn the Committee had received a stream of evidence from agricultural and conservation interests. In its report, the Committee welcomed what it perceived as 'a new mood' amongst farmers and their representatives, and reasoned that, with appropriate amendments to the Act and to agricultural policy, which went well beyond the necessarily limited measures of Clark's Bill, the voluntary approach could prove effective. These included a requirement for farmers across the countryside (and not just in SSSIs and the National Parks where management agreements have most commonly been used) to notify the conservation authorities of their application for an agricultural grant; provision to be made for National Park authorities to be able to apply for landscape conservation orders, analogous to nature conservation orders; greatly increased priority for conservation in the training and work of the Agricultural Development and Advisory Service (ADAS) staff; and greater resources for the negotiated purchase of threatened sites as an alternative to compensatory management agreements. The Committee prefaced these and other recommendations with the following warning:

> Our underlying concern is that, even if the changes we recommend are made to the Act and its administration, the wider agricultural structure will fuel the 'engine of destruction' . . . Without fundamental changes in the structure of agricultural finance, conservation will continue to be set in weak opposition to the forces of intensive and, paradoxically, frequently unwanted production, instead of being an integral part of good husbandry, as it should be. MAFF must reappraise its attitudes.

Reform of agricultural policies

The onset of the 'new mood' that the Select Committee on the Environment had noticed amongst farmers and their representatives can be traced back to a single event – the imposition, as we discussed in chapter 2, of quotas to cut milk overproduction, introduced by the European Council of Ministers in April 1984 to stave off the bankruptcy of the European Community (EC). The National Farmers' Union correctly interpreted this development as heralding the end of an era during which increased production was the be-all and end-all of agricultural policy, and immediately set about devising a new strategy. Conservation concerns, rather than being regarded as external or incidental, suddenly were seen to present a possible justification for continued financial support for farmers to offset the inevitable cuts in production incentives. The implications of milk quotas were not missed by other groups, and a wide-ranging debate on agricultural policy was opened up.

Shortly before the introduction of quotas, the Council for the Protection of Rural England had launched a *Campaign for the Countryside* aimed at securing 'fresh government agricultural policies more sensitive to conservation'. Representations were made to the Ministers of Agriculture and the Environment and to the environment ministers of the European Council. Support was also canvassed amongst MPs and peers. Writing to its county branches asking them to buttonhole their local MPs ('particularly Conservative backbenchers'), the Council for the Protection of Rural England's director advised them to stress that the initiative was 'thoroughly constructive . . . not "farmer-bashing" in any way, but placing the responsibility for desirable changes in the relationship between landscape/wildlife conservation and agriculture where it belongs – with the government's agricultural grant-aid policies' (6 April 1984). Talks were also initiated with the Country Landowners' Association and the National Farmers' Union to seek common ground.

A remarkable feature of the Council for the Protection of Rural England campaign was its links with the parallel one launched in April 1984 by *The Observer* 'to preserve our natural heritage and save Britain's countryside from further depredation'. The newspaper specifically urged that loopholes in the Wildlife and Countryside Act should be closed and that the Minister of Agriculture should adapt farm grants so as to promote conservation. Throughout the Summer and Autumn there followed a series of articles by their environment correspondent, Geoffrey Lean, highlighting specific threats to the natural heritage. Other newspapers, too, sensed a quickly ripening issue. The *Sunday Times*, for example, began campaigns in August 1984 to safeguard ancient woodlands and to control the use of pesticides.

The conservation agencies also felt emboldened to take a more public and sceptical stance towards government policies. In June 1984, the Countryside Commission issued a revised policy statement on *Agricultural Landscapes*, following consultants' reports demonstrating the marginal effect on the general deterioration of the farmed landscape of the Commission's efforts during the previous ten years. Sir Derek Barber, the Commission's Chairman, acknowledged that 'in a number of English counties, particularly in eastern England and the Midlands, the landscape is now only a shadow of its former character and beauty'. The statement shifted the emphasis of the Commission's policy in the lowlands from the development of new landscapes, towards the management and protection of those features, which remained. Despite this apparent admission of failure, the Commission continued to express its firm commitment to the voluntary approach of the 1981 Wildlife and Countryside Act, whilst urging that, to be given the chance to succeed, agricultural policy would have to be amended to discriminate in favour of activities which benefit the environment and against those which are detrimental.

The NCC was more forthright. Its report, *Nature Conservation in Great Britain* (26 June 1984), detailed 'the overwhelmingly adverse impact of modern agriculture on wildlife and its habitat in Britain'. Speaking at the launch of the report and its accompanying strategy, William Wilkinson commented forcibly, 'conservationists are regularly pressed to compromise. The answer in most cases must be no. As the review shows, there is little left with which to compromise. The salami-slicer has been at work too long.' Then with remarks clearly addressed to the then Secretary of State for the

Figure 7.4 Lowland landscapes under threat. Still a relatively intimate landscape of woodland and small fields, the Chilterns have been under pressure from intensive agricultural practices and urbanization – London is only 35 miles away. *Source*: S. and O. Mathews

Environment, Patrick Jenkin, who was present at the launch, and his Cabinet colleagues, Wilkinson urged that 'nature conservation needs a higher place in national priorities and a stronger claim on the nation's resources'. It was characteristic of the government's ambivalence that Jenkin, while welcoming the NCC's new strategy asserted that '*A miniscule part* of the sums of money that go to the support of agriculture and forestry diverted to the ends of conservation would work wonders' (editors' emphasis). However feebly, this response did at least concede in principle the major point on which there was agreement across the whole conservation movement, from the Countryside Commission to Friends of the Earth – that part at least of the agricultural budget should be redirected to support conservation-orientated husbandry.

An inquiry by a House of Lords Select Committee into draft regulations from the European Commission on improving the efficiency of agricultural structures provided an unexpected platform for conservationists to publicize their arguments for the reform of agricultural policy. The draft regulations did little but tinker with the existing framework for agricultural grants, even though the EC's Third Action Programme on the Environment, agreed by the Council of Ministers in 1982, stated the need to integrate environmental objectives into agriculture.

Throughout the Winter and Spring of 1983–84 a string of conservation witnesses paraded their arguments before the Lords Committee. With unprecedented unanimity the Countryside Commission, the NCC, the Royal Society for the Protection of Birds, the Council for the Protection of Rural England, the Council for National Parks and others argued the case for drastic changes in agricultural support, away from maximizing production and towards the integration of food production with wildlife and landscape conservation. In response, MAFF's civil servants emphasized the regard they already showed environmental considerations, but they stuck firmly to the traditional MAFF line that environmental benefits must always be incidental to the central purpose of agricultural aid – namely, farming 'efficiency'. They claimed that this was also the correct legal interpretation of the Treaty of Rome. Brian Peart, Under-Secretary at MAFF, argued that unless government schemes for agriculture and for environmental improvement were kept separate, 'you are in danger of having confused objectives and . . . expensive administration.' In the event, the select committee accepted neither of these points in its final report published in July.

MAFF and the DoE were criticized for adopting 'an unnecessarily narrow attitude to the Treaty of Rome'. Contrary to their representations, the Committee proposed that 'care of the environment should have comparable status with the production of food' in the promotion of farming improvement. Far from endorsing a sharp demarcation between environmental and agricultural policy, the Select Committee called for 'the revision of existing priorities and greater co-ordination and co-operation' between the two departments. 'In the past', it complained, 'the DoE have been largely subordinate to MAFF, and have not been active enough in promoting care for the environment'. The DoE was urged to 'revise their role in relation to agriculture generally', and MAFF was berated for being 'backward-looking' and too 'production-orientated'. Both departments were roundly upbraided for being insufficiently responsive to the strength of public opinion on the countryside.

Such a swingeing report indicated the extent to which conservation groups had captured the high ground in the conflict with the farming lobby. When the House of Lords debated the reports on 23 July 1984, Lord Belstead, the then Minister of State for MAFF, announced that the government had decided to seek a completely new title in the EC structures regulation 'conveying powers which would enable us in environmentally sensitive areas to encourage farming practices which are consonant with conservation'. He claimed that this would 'herald a totally new policy for balancing agricultural and conservation objectives'.

The different responses of the National Farmers' Union and the Country Landowners' Association to this development are instructive, and highlight the different styles of the two organizations. The National Farmers' Union briefing for the Lords debate warned that 'British producers must not be placed at a competitive disadvantage vis-à-vis their continental counterparts', and concluded: 'although there is scope for taking more account of environmental objectives in the regulations, the encouragement of a thriving agriculture must remain paramount.' In October, the National Farmers' Union issued its *New Directions for Agricultural Policy* which gave priority to ensuring the financial positions of the industry, while calling for 'new integrated policies . . . to cover the whole complex of farming, rural development and environmental needs' in the uplands and, more generally, for encouragement to farmers 'to take proper account of environmental needs'.

In contrast, on the eve of the Lords debate, the Country Landowners' Association joined with the Council for the Protection of Rural England for the first time in calling for 'changes in policy which would help end the unfortunate and damaging conflicts over the effect of modern agricultural practices on the countryside'. The joint statement continued:

> we are determined to work urgently together to obtain adjustments to agricultural support policies, which will establish a better balance between efficient agriculture, private landownership and the public interest in conservation and enjoyment of the countryside.

The Country Landowners' Association then set up a working party to formulate policies for integrating agriculture and conservation. As its report made clear when published in September, the rationale was to ensure that the money being lost to farming was not lost to the countryside. The Country Landowners' Association's report went much further than the National Farmers' Union in recognizing the desirability of redirecting public funds where they were most needed in both social and environmental terms, and in accepting the case for planning controls over farm buildings and roads.

The Halvergate controversy and environmentally sensitive areas

While the failings of the Wildlife and Countryside Act and the excesses of the farm support system were being widely debated in the press and Parliament, the contradictions between conservation and agriculture were starkly revealed in a succession of local controversies. These typically concerned a valued terrain threatened by large-scale, government-promoted land improvement. In the uplands, this might be the ploughing up of moorland or afforestation; in the lowlands, the reclamation of wetlands, marshes and water meadows facilitated by arterial drainage and river improvement schemes. The size of the areas involved and the sweeping pressures they faced tested the ability and resolve of conservation authorities to pursue their responsibilities. Also revealed was the large and open-ended commitments, under the 1981 Wildlife and Countryside Act, to compensate farmers in such circumstances for refraining from damaging the environment. The fact that this compensation had to be paid from the meagre budgets of the conservation authorities to dissuade farmers from taking advantage of MAFF-financed improvement schemes pointed up the inequity and irrationality of

Figure 7.5 The grazing marshlands of Halvergate. Norfolk wetlands were at the centre of the controversy over the 'improvement' of agricultural land that led eventually to the development of the principles underlying Environmentally Sensitive Areas.
Source: Countryside Commission

these arrangements, even more so, when the improvements promised merely to add to food surpluses.

The local controversy which proved to be the most tortuous and intractable was over threats to drain and convert to arable land the Halvergate Marshes in the Broadlands of Norfolk – the last remaining extensive stretch of open grazing marsh in Eastern England. Here the policy contradictions became quite unmanageable and Ministers were forced to fly by the seat of their pants in an effort to rescue their conservation policies from ignominy. Through a series of *ad hoc* responses to a succession of crises, a new approach emerged to resolving the conflicts between conservation and agricultural interests in this and other pressured areas.

During the early 1980s with the loss of the already much depleted grazing marshes in the Broads running at more than 5 per cent per annum, the Broads Authority looked to the provisions of the 1981 Wildlife and Countryside Act to safeguard what was left. The gains to be achieved from conversion to highly subsidized cereal crops were such that compensation of between £250 and £400 per hectare per annum would have been needed. Over a five-year period, with perhaps 2000 hectares coming under threat, compensation payments could rise to £1 million annually. The Authority was entitled to 50 per cent support from the Countryside Commission, but still the cost of management agreements to save Halvergate seemed prohibitive.

At a meeting in March 1984 between the Broads Authority, the Countryside Commission, MAFF and DoE Ministers, the latter demurred at a request that the Broads be given 90 per cent grant-aid from central funds. They did, however, agree to consider the other proposal made, that MAFF act to increase the effective revenue from livestock. A working party was set up to make proposals and in the meantime the Broads Authority sought to stop four landowners from ploughing for a year by offering a token payment. David and Michael Wright, joint owners of 150 hectares refused, however, giving the Authority until June 10 to reach an agreement.

The working party could not agree, but deadlock was broken by an ingenious Countryside Commission proposal whereby it undertook to finance an alternative system of livestock support payments over a three-year period in an attempt to sustain the traditional grazing regime on the marshes.

The Commission stipulated that MAFF should subsequently take over the support payments should they prove effective, but when the working party reported in mid May there was, as yet, no ministerial agreement on future financing. The Authority sought a one-year agreement with Michael Wright whose land was critical to the landscape value of the core marshes, but decided not to negotiate with his brother who, on June 10, duly began to prepare his ground for ploughing only to be thwarted by direct action from the local Friends of the Earth. He subsequently sought to negotiate, but agreement could not be reached and ploughing recommenced.

Another landowner, some distance from Halvergate, who seemed wholly uninterested in any form of management agreement, also indicated an intent to plough. The publicity attracted by this case caused Ministers acute political embarrassment. The Broads Authority and the Council for the Protection of Rural England successfully encouraged the Secretary of State for the Environment to use an Article 4 Direction to prevent the drainage (an Article 4 Direction effectively places the designated works under the development control system). Its imposition simultaneously exposed the voluntary philosophy underlying the 1981 Wildlife and Countryside Act and delivered a body blow to the argument that planning controls are wholly inappropriate in regulating agricultural development. The government would not, however, countenance a 'blanket' Article 4 Direction over all the Broads grazing marshes which Friends of the Earth and the Council for the Protection of Rural England lobbied for and Ministers faced a barrage of hostile Parliamentary questions throughout July as the activities of David Wright caused continued outrage.

Although the Government could point to the refusal of the Broads Authority to seek a management agreement, it could not but recognize that the Broads Authority's evident inability to finance many management agreements indicated that other means of protecting this highly valued landscape must be found. Now implicated in the fate of the marshes, Ministers were anxious to find a lasting solution.

The height of the Halvergate controversy coincided with the debate over the draft proposals from the European Commission to revise the structural supports for agriculture. As we have seen the House of Lords Select Committee which reviewed the latter was highly critical of the failure of the DoE and MAFF to reconcile agricultural and environmental policies. When the Lords debated the Committee's Report, agriculture ministers announced

Figure 7.6 Halvergate under the plough. The loss of traditional marsh grazing land in this most sensitive part of the Broads landscape served visibly to reinforce the deficiencies of the 1981 Wildlife and Countryside Act.
Source: R. Denyer

Figure 7.7 Public protests at Halvergate, 1984. Whilst the Broads Authority attempted to put a temporary stop on further marshland ploughing, local conservationists were looking to the government for a more permanent solution. *Source: Eastern Daily Press*

in both Houses that the government intended proposing an amendment to the revised EC structures directive which 'would be designed to allow member-states to pay aids to farmers in suitably designated areas of high conservation value in order to encourage farming practices beneficial to the environment'. By the Autumn of 1984 Britain had succeeded in convincing the Commission of the European Communities and other Member States of the need for this innovation, and so the first common environmental aid scheme was introduced. Article 19 of the new Structures Regulation enabled agriculture departments of Member States to designate areas where 'the maintenance or adoption of particular agricultural methods is likely to facilitate [the] conservation, enhancement or protection of the nature conservation, amenity or archaeological and historic interest of an area', and to give financial incentives to encourage appropriate farming practices in these Environmentally Sensitive Areas (ESAs).

Meanwhile the Countryside Commission had proceeded to set up its experimental scheme to protect the Broads grazing marshes. It was designed to offset some of the economic pressures facing traditional livestock graziers in the area. The Broads Grazing Marsh Conservation Scheme offered farmers a flat rate annual hectarage payment in return for an agreement to abide by certain constraints on stocking densities and the use of fertilisers and herbicides. In March 1985, MAFF announced that it would cover half the costs of the scheme which was to be jointly administered by the Ministry and the Countryside Commission. Covering almost 5000 hectares, the grazing scheme attracted the participation of over 90 per cent of the eligible land. A measure of its success is that no marshland has been converted since 1986. Its wider significance, though, has been as a prototype for the ESAs including the principle of flat rate subsidies for traditional husbandry rather than the individual 'profits foregone' arrangements enshrined in the 1981 Wildlife and Countryside Act.

In February 1985, MAFF issued a discussion paper on the introduction of ESAs. It was proposed that they should cover areas whose national environmental significance was threatened by agricultural change, but which could be conserved through the adoption or maintenance of a particular form of farming practice. The areas should also represent a discrete and coherent unit of environmental interest to permit the economical administration of appropriate conservation aids.

The Countryside Commission and the NCC were asked, in consultation with English Heritage and its Welsh counterpart, to recommend a shortlist of possible ESAs. First they separately drew up comprehensive lists of candidates areas – 160 for the NCC and 150 for the Countryside Commission. Using the above criteria, the lists were reduced to 56 and 46 respectively. With pressures to make rapid progress on a first tranche of designations, and with MAFF indicating that financial restrictions would mean initially only a few ESAs, the two conservation agencies agreed a priority shortlist of 14 areas, that would benefit from early designation and would give a reasonable geographical spread:

Anglesey	Pennine Dales
Brecklands	Radnor
Broads	Somerset Levels and Moors
Cambrian Mountains	South Downs
Clun (Shropshire Borders)	Suffolk River Valleys
Lleyn Peninsula	Test Valley
North Peak	West Penwith

In two rounds of designations – announced in mid-1986 and mid-1987 respectively and introduced early the following years – all of these suggestions were taken up by Ministers, except for Radnor and Anglesey (reflecting a more guarded response to the ESA concept on the part of Welsh agricultural interests). In addition, the Mountains of Mourne in Northern Ireland, and Loch Lomond, Breadalbane, the Stewartry of Kirkcudbright, the Uists and Whitlaw/Eildon in Scotland were also designated. The total area covered by ESA designations was 738,000 hectares.

The areas are very different in character: from the lowland heaths and grassland of Breckland and the wet pastures, reedbeds and willow-lined rivers of the Suffolk valleys; to the rounded hills and steep, wooded valleys of the Shropshire borders and the open moors of peat and heather of the North Peak. What they have in common, though, is that their landscape and ecology depend critically on the continuation of agricultural systems which are typically, extensive and livestock based, though some are threatened more by decline and others more by improvement and intensification.

Following its experience with the Broads Grazing Marsh Conservation Scheme, the Countryside Commission proposed that management prescriptions and incentives should be simplified as much as possible. This meant payments at a standard rate per hectare in each area. In some areas, though, the complexity of landscape types and habitats demanded a second tier of hectarage payments, associated with additional constraints. The agricultural requirements of the agreements were formulated by the agriculture departments upon the suggestions of the Countryside Commission and the NCC and following local consultations.

They vary according to the physical and farming circumstances of each ESA. Most include restrictions on fertiliser use and stock densities, prohibitions on the use of herbicides and pesticides and on the installation of new drainage or fencing. Farmers also agree to manage appropriately features such as hedges, ditches, woods, walls and barns and to protect historic

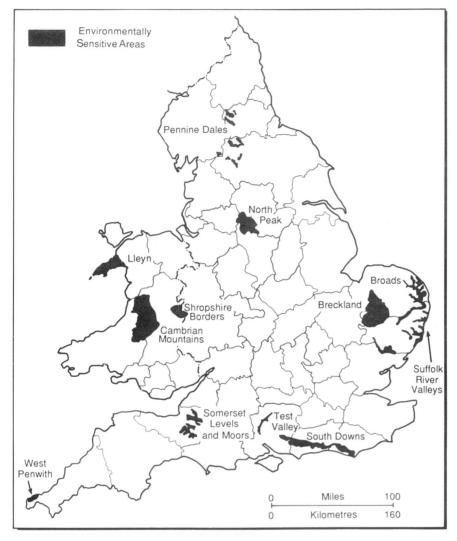

Environmentally
Sensitive Areas

Pennine Dales

North
Peak

Lleyn

Broads

Shropshire
Borders

Breckland

Cambrian
Mountains

Suffolk
River
Valleys

Somerset
Levels
and Moors

Test
Valley

South Downs

West
Penwith

| 0 | Miles | 100 |
| 0 | Kilometres | 160 |

Figure 7.8 The Environmentally Sensitive Areas of England and Wales. These new designations have increased the variety of our protected landscapes and provide a much more effective approach to conservation.
Source: Countryside Commission

features. Each ESA is administered by a local ADAS officer advised by a steering committee of conservationists. Although it is far too early to gauge the environmental results, the response of farmers has been good. By early 1988, in England alone, over 1,400 farmers had applied to join the first round of the scheme, putting forward some 32,000 hectares which represented 78 per cent of the land that the ministry considered suitable for inclusion. The government's budget for ESAs nationally was running at about £12 million per annum.

The ESA programme, with its direct and tangible commitment to conservationist farming, marks a significant departure for agricultural policy. For the first time farmers are being paid by the agriculture departments to 'produce' countryside; this perhaps is, the most radical feature of the ESA programme, distinguishing it from previous conservation measures administered by the conservation agencies. Moreover, in overcoming one sectoral division it has leap-frogged a second – that between landscape protection and nature conservation – which has become an increasingly anachronistic feature of British conservation. It has also created important precedents in the deployment of agricultural funds and personnel. A 1987 amendment to

Figure 7.10 'Now lads, who's going to ask the Minister about ESAs?'
Source: Farmers' Weekly

the European Commission's structurse regulation also recognized the significance of ESAs as an instrument of farm adjustment, and it provided for an element of Commission support for ESA payments.

Nevertheless, calls to extend the ESA approach more widely across the countryside have been ignored by MAFF. The ESA programme, therefore, remains a minor though well-publicized aspect of policy. This has led some environmentalists to regard it as a distraction from the main task of reforming agricultural policy more generally. Even the Countryside Commission has argued that altering the administration of Hill and Livestock Compensatory Allowance (headage) payments, so that they no longer encouraged over-stocking and grassland improvement, would greatly diminish the need for ESAs in the uplands. Another legitimate criticism lies with the voluntary nature of farmer participation. While it is quite appropriate that such a novel scheme should seek to foster the confidence and co-operation of farmers and not alienate them, it could be argued that this is being taken to excessive lengths. For example, a farmer can enter just these parts of his farm and be paid accordingly, but still be free to intensify production on the rest. Likewise. those farmers who choose not to participate in the ESA programme can undertake environmentally damaging improvements and may even be eligible for MAFF capital grants.

The agricultural counter-response

ESAs must be seen as part of a much wider effort by the agricultural lobby to regain the initiative in the continuing debate over the future of the country-side. They were introduced into British legislation by the 1986 Agriculture Act. Unlike the many earlier Acts of the same title, this one contained measures as we noted in chapter 1 which positively tackled matters of central importance to conservation. The Act also imposed a statutory duty on Agriculture Ministers to balance the conservation and promotion of the enjoyment of the countryside, the support of a stable and efficient agricul-tural industry, and the economic and social interests of rural areas.

The relevant clause introducing this duty which we cite in chapter 1, had been tabled by the government in January 1986 during the Bill's committee

Figure 7.9 Environmentally Sensitive Areas – a European model? ESA designations in England and Wales, such as that covering the Somerset Levels, have been accepted by the European Community as a means of reconciling agriculture with conservation.
Source: Countryside Commission

stage. Only nine months earlier, it had deleted a clause with similar wording in Dr David Clark's Bill. Likewise Ministers had resisted previous efforts during the passage of the Wildlife and Countryside Act. How can we explain this U-turn? First, it represented a triumph for the steady attrition by the conservation lobby which had built up broad backbench support for the reform. But it also marked a major shift in strategy by MAFF, which had a reputation for being inward-looking and defensive in protecting its own departmental territory and the interests of the agricultural sector. No longer was this a tenable strategy, with so much critical debate surrounding agricultural policy, with Thatcherite pressures to commercialize, privatize and reduce many of its traditional functions, and with speculation in the farming press extending even to MAFF's possible demise. In redefining its broad responsibilities the Ministry went some way toward embracing the proposal the Country Landowners' Association had been pursuing for 10 years for a Ministry of Rural Affairs. With the change, the Ministry adopted a much more active and promotional stance generating a flow of policy initiatives and assuming a more positive role in the promotion of conservation in the farmed countryside.

One of the first signs of the change in direction, apart from the ESAs, occurred with the launch in 1985 of MAFF's new scheme for agricultural grants to replace the farm improvement schemes, which had been so roundly condemned for encouraging environmental degradation. The new scheme stipulated that plans should take into account the impact on the countryside of improvements. For the first time MAFF was formally empowered to refuse grants if proposals were considered environmentally detrimental. In what was also an exercise in slashing direct supports to farmers, a number of types of capital improvement grants were either curtailed or scaled down. But grants were to be available for planting hedges, repairing traditional walls, planting broadleaved shelter belts and even for employing the services of specialist consultants to provide landscaping advice for farm development schemes.

Two of the central components of the agricultural counter-response deserve our special attention: the promotion of the county Farming and Wildlife Advisory Groups (FWAG) and the encouragement of alternative land uses in the face of continuing growth and agricultural surpluses.

FWAG was formed in 1969 as a national forum for conservationists and agriculturalists to meet together to foster mutual understanding and co-operation. In the early and mid 1970s it sponsored a series of practical farming and wildlife exercises designed to demonstrate the compatibility of farming and conservation practices. These were usually located on success-ful, commercial farms which had devoted spare resources to tree-planting, pond creation or the management of what marginal habitats remained. Until the late 1970s it was a rather marginal ginger group dependent mainly upon the financial backing of the Royal Society for the Protection of Birds.

In the controversy surrounding the passage of the 1981 Wildlife and Countryside Act, much attention was devoted to the majority of the farmed countryside not specifically covered by the Act. Agricultural leaders and ministers were quick to emphasize that the principles of goodwill and voluntary co-operation embodied in the Act were equally applicable to the wider countryside. In particular the rapid emergence of county FWAGs in the late 1970s and early 1980s was perceived as clear evidence of a concern for conservation in the farming community. Their establishment and growth were fostered by the National Farmers' Union, the Country Landowners' Association and ADAS. Since then, these FWAGs have been presented as the best available vehicle for demonstrating the capacity of farmers and

Figure 7.11 Information from Farming and Wildlife Advisory Groups. The voluntary principle has been enhanced by a comprehensive system of general information available to the agricultural community at large.
Source: D. Noton

conservationists locally to work together in harmony and as a means by which farmers themselves might be encouraged to adopt conservation practices. Their avowed principle is that conservation and modern farming need not be incompatible and that loss of wildlife habitats through agricultural intensification can best be ameliorated by encouraging farmers to modify their practices through the provision of appropriate advice and encouragement.

FWAG is the prime expression of the voluntary principle in conservation. This has two components. One is, positively, to stimulate and broadcast amongst farmers and landowners a social ethic concerning stewardship of the countryside, including the protection and enhancement of natural diversity and beauty within the context of modern farming practice and estate management. The other component is an ideological defence of the autonomy of farmers and landowners from statutory controls, though an emphasis on the paramount need to retain their goodwill and voluntary co-operation if workable remedies to conservation problems are to be found.

Both these components have been pressed into service by the agricultural lobby.

If county FWAGs were to play a self-regulatory role, however, it was important that they did more than preach to the converted. They needed to develop an advisory service that could be available to all farmers and landowners. This called for professional extension staff. In February 1984, in what represented a major triumph for FWAG, a Farming and Wildlife Trust was launched principally to raise the money needed to fund the employment of 'farm conservation advisors' in the counties. It was planned to appoint up to 30 advisors over the following two to three years and to this end an appeal target of £500,000 was established. Significantly, half of this amount had already been pledged, in advance of the formal launch, by a variety of charities, countryside agencies and interest groups, agricultural companies and prominent farmers and landowners.

By far the biggest sponsor was the Countryside Commission which, as well as granting £40,000 towards running and establishment costs for national FWAG and the new trust, also had agreed to match the funds raised by the trust by meeting half the cost of each advisor for a period of three years. Over the next ten years the Commission was prepared to commit £1 million to the venture. Much assistance in kind was to be given by MAFF, which agreed in many counties to supply secretarial support for the groups. ADAS officers were encouraged to assist local FWAGs and MAFF cited its involvement with FWAG as demonstrating its own commitment to conservation.

Between 1984 and 1986 the majority of English counties and several Welsh ones appointed full-time FWAG advisors. Their task is not an easy one. He or she (the majority are in fact female) is required to demonstrate both farming and conservation knowledge and expertise. The advisor, often a young graduate, is usually expected to undertake a strenuous round of lecturing and publicity work on top of the priority task of advising farmers on any aspect of conservation on farms.

The enthusiasm and energy of the advisers is much to be applauded.

Figure 7.12 The FWAG personal touch. County based representatives, as well as providing printed materials, tailor advice to the special needs of individual farmers. *Source*: Countryside Commission

However, one adviser in a county can only expect to reach a small proportion of the farmers, even over a five-year period. Moreover, they can only respond to requests for advice and then must tailor the advice they give to the needs and circumstances of the farmer. They have been criticized for offering only cosmetic advice and for not tackling fundamental problems of agricultural change. This is not entirely fair for they are dependent upon the goodwill of farmers and have few incentives available by which they can persuade farmers to go beyond fairly minor adjustments to farming systems.

ADAS's supportive role for local FWAG groups has been brought into question by the increasing pressure it has come under to take a lead in the provision of conservation advice. This was highlighted in the review conducted by its incoming Director-General in 1984 and even more strongly in a House of Commons Agriculture Committee report on the Service in 1985. The 1986 Agriculture Act, also placed upon MAFF the responsibility to charge for ADAS advice. In debate the Minister gave assurances that exceptions to the charging rule would be made for advice on conservation, farm diversification and animal welfare.

In March 1986 an Environmental Unit was established to give a co-ordinating lead to ADAS on conservation issues. Then in November 1986 the Agricultural Service was merged with the much smaller Land and Water Service (LAWS) to form a new Farm and Countryside Service, a change in emphasis and tactically in name too, that we noted in the Preface to this volume. The significance of this lies in the fact that hitherto conservation advisory duties had resided with LAWS (comprising specialist estate management and surveying expertise) and was somewhat marginal to the overall work of ADAS. Now, however, there is no organizational separation between the services providing agricultural advice and conservation advice. ADAS officers accept that a major element of their work in the future will lie in advising on the implementation of conservation policies.

In assessing ADAS's capacity to perform the new role allotted to it, lessons from its past need to be noted. The post-war history of the service shows a preoccupation with narrowly defined production issues and a retreat from a service aiming to supply advice to any individual farmer seeking it, to that of a general information service of greatest use to the larger and more sophisticated ones. As a result many farmers have had little or no direct contact with it. By the same token ADAS officers have not developed the complex educational skills required in other countries where the 'farm problem' has required an advisory service with an active rather than a responsive role. It is very clear that, in contrast, the promotion of conservation policy will require an active and innovative approach capable of reaching and influencing the majority of farmers.

The 'commercialization' of other aspects of ADAS's work will not help matters: if an adviser's visit does not take place for financial reasons, then the opportunity to offer conservation advice will be lost. In addition, ADAS must now compete with the less than impartial services already offered by manufacturers of pesticides, chemicals and farm machinery. The fact that conservation advice remains free may encourage ADAS to expand this service, especially if demand for other services, for which payment must be made, becomes slack. However, it is seriously constrained by staff cuts, lack of suitably trained advisers and minimal provision for retraining or recruiting specialist personnel.

The closeness of the links between ADAS and FWAG mean that FWAG is not insulated from these difficulties. However, it has a number of potential advantages over ADAS. It does not have to retrain agricultural staff and has been able to recruit specialist advisers who are usually extremely well qualified in ecology and conservation. As a fresh organization in the advisory field it has a pioneering vigour not present in ADAS. Moreover at a time when ADAS's links with the farming community appear to be weakening, many county FWAGs seem to be developing broad grass-roots support.

If there is to be a struggle between ADAS and FWAG, rather than the partnership of recent years, it is unlikely to be an open one. MAFF is not likely to advocate the disbandment of FWAG. Instead it will seek resources to develop its own farm conservation advisory service and resist the expansion of FWAG's extension capacity. It is likely to promote instead FWAG's role as a forum and seek to re-direct the FWAG advisory effort to a more general promotional and information giving role. There are already indicators in busy counties that FWAG officers are forced into directing farmers to more specialist sources of advice.

A crucial factor in the development of the respective roles of FWAG and ADAS will be the extent to which conservation policy for agriculture extends beyond that of advice and education. FWAG is likely to be at its strongest if policy remains in this sphere. A more directly regulative policy, or one dependent upon grant inducements, would require the more direct governmental involvement of ADAS. An extension of the ESA principle is a case in point as are various proposals to make grant aid conditional upon conservation-sensitive farm management plans.

Increasingly, new measures for farm adjustment are being introduced to tackle over-production and explicitly or implicitly they promise indirect conservation benefits. Indeed, in the case of proposals to leave farm land idle or to de-intensify production, conservation has been presented as part of the rationale. However, much depends on the details of implementation and the extent to which conservationists can secure stronger regulatory mechanisms. The Farm Woodlands Scheme, introduced in 1988, for example, might easily lead to plantings of commercial species with very limited conservation benefit and even losses if planting takes place on semi-natural areas. Similarly a set-aside scheme, especially aimed at reducing cereals production, may well prove disappointing for conservationists.

The extensification scheme promises reduced production for which farmers are to be compensated, the opportunity for new enterprises and environmental benefits. The framework for such a scheme was established by the European Community Regulation of June 1987 which defined extensification as involving the withdrawal of at least 20 per cent of a farm's productive capacity, where the 'land may be left fallow with the possibility of rotation, afforested or used for non-agricultural purposes'. In addition to the central aim of reducing surpluses, MAFF in its consultation document on the scheme, highlights three other desirable objectives all of which are consistent with the Council regulation. First, there is the freeing of some land from intensive agricultural production to be used for purposes such as the growing of trees, conservation and amenity, and new farm enterprises. Secondly, and more generally, there is the fostering of a changed pattern of land use in the countryside so as to improve conditions for wildlife and to enhance the landscape and the environment generally. Finally, there is the adoption of administrative rules which are, within the constraints of EC requirements and the need to safeguard public funds, as simple as possible so as to avoid burdens and costs on applicants' businesses.

The first two aims are ambitious ones, but without special incentives to ensure their success, doubts must be expressed as to whether a deliberately unobtrusive scheme will be able to achieve these broader objectives. By insisting on the voluntary approach MAFF is ignoring the advice of the

National Farmers' Union, which is convinced that only a compulsory scheme will be seen as fair and workable by farmers.

In failing to provide adequate details of how the conservation objectives might be met the Ministry appears to have disregarded the views of the Countryside Commission given three months before the publication of the consultation document. The Commission argued that set-aside land would need well considered management and that to maximize its value, farmers should be encouraged to take out of production the land with the greatest conservation or recreation potential, and preferably in schemes which bring together blocks of land on adjacent holdings, to create significant landscape features and wildlife habitats. The Commission also called for less intensive management of the land remaining in production and arrangements to prevent unimproved land, often of high conservation value, being taken into more intensive cultivation to compensate for the effect of set-aside on farm output.

Although the MAFF consultation document mentions some of these aims, its proposals for implementation do not instil confidence that the Countryside Commission's concerns are being seriously considered. In particular, the document opts for a reduction of the area on a farm devoted to cereals so leaving open the possibility of intensification of production on the remaining land with adverse environmental consequences. Management decisions for fallow land remain in the hands of farmers and a Code of Practice is proposed to encourage farmers to manage land for conservation purposes, but subject to no special inducements or environmental monitoring. There is no mention of the involvement of conservation interests either in the selection of land suited for set-aside or its management. Thus although publicized by the Ministry, as part of its new-found commitment to conservation, the scheme is

Figure 7.13 The Flow Country of Sutherland. Commercial timber development, until recently spurred on by tax concessions for the rich, has proved controversial in one of Britain's few remaining primaeval landscapes. *Source*: Forestry Commission

unlikely, unless amendments are made, to achieve its more noteworthy objectives.

Forestry developments

Alongside the changes in the agricultural sector, a somewhat quieter re-orientation has been taking place in forestry. As we noted in chapter 3, the Forestry Commission has long been required to pay heed to questions of landscape and public access. However, until recently, less attention has been paid to specific questions of wildlife conservation. A number of recent developments have served to change both the focus of concern regarding forestry and the political complexion of forestry developments. Until the early 1980s, upland afforestation was more likely to be disputed on landscape than on ecological grounds. In the National Parks at least, a measure of reconciliation of planting with landscape considerations has been achieved. Wildlife concerns surfaced in a few cases, such as the ornithological implications of afforestation of the Berwyns in the early 1980s. Under the Thatcher government, the role of the Forestry Commission in direct planting has been much reduced and that of the private sector expanded. Timber Growers UK, following the lead of the agricultural lobby, issued a code of practice on afforestation in an attempt to forestall criticisms of private forestry companies.

In recent years the conflict between afforestation and conservation has emerged more sharply with regard to the planting of large areas of the 'flow country' of Sutherland and Caithness. A unique expanse of blanket bogs of international botanical and ornithological importance, it had hitherto escaped the attention of foresters due to problems associated with drainage and windthrow. However, in their pursuit of fresh outlets to reap the tax benefits of afforestation some of the major forestry companies have been prepared to risk planting in this area. It remains to be seen how the abolition, in the 1988 Budget, of some of the tax concessions which favoured afforestation will affect disputes in the 'flow country' and elsewhere, especially as the change was accompanied by a sizeable increase in the grants for forestry development.

The importance of the private forestry sector means that the Forestry Commission itself is not so directly and negatively implicated in the conflict between conservation and afforestation as formerly. The Commission's task regarding upland afforestation is now the provision of grants to companies such as the Economic Forestry Group and Fountain Forestry, as well as to private landowners. Grants are not mandatory and local authorities, the NCC and other bodies are consulted on contentious grant applications. With the shift in emphasis from tax concessions to grants, the regulatory role of the Commission is likely to attract greater scrutiny. The role of county and regional authorities too, is likely to become more significant, as we have argued in chapter 3.

In recent years, the Forestry Commission, like MAFF, has devoted much energy to re-defining its role partly in conservation terms. Its Broadleaved Woodland Grant Scheme, launched in 1985, marks the main development here. A remarkable feature is that timber production is not required to be the main objective of a planting proposal. Conservation and amenity are equally valid objectives. However, the scheme is not a panacea for conservationists. Some would maintain that commercial deciduous monocultures may be as environmentally destructive as conifers. Concern has been expressed at the possible consequences of planting certain highly competitive species, such as beech or sycamore, alongside existing mixed woodland. There have even been allegations that conifers have been planted under the scheme by a re-interpretation of the grant rules.

The Forestry Commission, therefore, finds itself in the late 1980s in a very similar position to MAFF. It has attempted to re-assert its role in the policy arena by adopting conservation orientated policies, though these have not entirely placated conservationist critics. One striking parallel with regard to the imbalance between rhetoric and reality is the implications of the ideology of privatization on each organization. MAFF has made much of its role in providing conservation advice to farmers, but at the same time its advisory service has been reduced in size and made commercial in function. The Forestry Commission emphasizes its promotion of conservation and amenity at a time when it is being forced to sell some of its woodland to the private sector.

Still room for improvement

The 1980s have undoubtedly been years of tremendous political activity in rural conservation. The conservation movement has emerged as a significant force enjoying widespread popular support. A number of major achievements can be recorded. Across the National Parks and SSSIs, conservation-orientated land management is generally now the established practice, yielding important amenity and wildlife benefits and contributing to the support of the rural economy and society. Meanwhile, in the ESA programme, the important principle has been conceded that the production of countryside is a legitimate objective of agricultural policy and the disbursement of agricultural funds.

But in other respects, progress has been disappointing. Even in these highly cosseted areas, the principal achievement has been to slow down the rate of destruction. The NCC in its annual report for 1986–87 notes that 'for the second successive year, we can report that no SSSI notified under the 1981 Wildlife and Countryside Act has been lost. Two sites, which had been approved as being of SSSI quality and which were awaiting notification, were lost during the year and 46 SSSIs suffered some lasting damage.' The NCC expresses due concern at these cumulative losses, but hopes that they will be further reduced as the notification programme nears completion. This essentially rearguard action through the cumbersome machinery of the 1981 Wildlife and Countryside Act has effectively consumed the bulk of the energies of the NCC's regional staff for most of the decade. Inevitably, its general advisory role and responsibility for the wider countryside have suffered.

In the wider countryside, the evidence we have is of continued indeed accelerated, environmental destruction. There is, as we noted in chapter 2, the growing problem of agricultural pollution with increased incidence of farm wastes and the contamination of water courses with agricultural run-off. Recent survey work by the Countryside Commission and the DoE indicates an increased rate of landscape loss, with, for example, the average annual rate of hedgerow removal in England and Wales rising to 4000 miles between 1980–85, compared with 2,900 miles between 1969–80. In 1988 there is little evidence to indicate any change in the rates of removal identified in the early 1980s.

A measure of the true worth of the new conservation responsibilities of MAFF and ADAS will be the extent to which these dismal trends are halted and reversed. Clearly, a major task for the conservation movement is to ensure that MAFF pays more than lip service to conservation. While the recent broadening of its responsibilities is to be welcomed, there is a sense here, too, that the 1980s have been wasted years, since just such a reform was proposed back in 1978 by MAFF's own official advisory council in the 'Strutt Report'. Even so there is the risk that the change will be merely symbolic. There is no tangible evidence, for example, that MAFF had previously taken seriously the requirement on all government departments, under the 1968 Countryside Act, to 'have regard to the desirability of conserving the natural beauty and amenity of the countryside'. The lack of serious attention and specific commitments to conservation in recent proposals for extensification and land diversion certainly do not indicate a major re-orientation in this Ministry's outlook.

The question arises then of what became of that coalition of forces which, in the early 1980s, seemed poised for sweeping reforms to agricultural policy. For a start, the neo-liberal critique of farm policy became marginalized within the Conservative Party, as Ministers realized the difficulties of reforming the CAP without jeopardizing the future of the EC. The Government also refrained from too severe a squeeze on Britain's farmers out of concern for the electoral consequences.

The second factor is that rural conservationists both voluntary and official, regard rural landowners more than other interest groups as their natural allies. In part this may reflect an incipient anti-urbanism in their outlook but, in addition, through many years of having to collaborate with farmers and landowners as the owners and controllers of the countryside, they have grown particularly attentive to their interests. They have therefore been reluctant to throw themselves behind a radical reform of agricultural support, preferring instead to seek concessions for conservation as part of a process of constructive reform. Undoubtedly, questions of political pragmatism have also been a consideration. Certainly, those groups, most notably Friends of the Earth, which have pressed for more sweeping cuts in support prices and much stronger environmental controls seem to have had less impact on the course of events than those that have allied themselves more closely to the land-owning interests.

Figure 7.14 'Its Nicholas Ridley frightening the birds with his privatization plans.'
Source: *Courtesy of Geoffrey Dickinson, Financial Times*

Chapter 8

New rural policies to the turn of the century: is the sum greater than the parts?

A rural dichotomy

In Mrs Thatcher's first government Michael Heseltine was in charge of the Department of the Environment. At that time he was advocating policies in the countryside which seemed to push economic development to the forefront of the rural agenda with environmental issues a secondary consideration. Indeed, in 1980 he declared that his strategic land-use planning guidance was designed to 'sweep away obstacles to commercial enterprise' in the rural South East of England. And yet in the early Spring of 1988 he was able to write a letter to the current holder of that office whose views, one might imagine, ought not to be far removed from his own. We reproduce this as Figure 8.1 which you should read now.

It is instructive because it underlines a clear-cut dichotomy about our countryside. Heseltine as a Minister of the Crown naturally had to espouse policies which related to the government's view of the broad interests of the nation in terms of its countryside, a task which Nicholas Ridley inherited in 1986. It was Heseltine who, with a wider concern for meeting the housing needs of the South East told Berkshire County Council that its structure plan for the central area would have to be revised to accommodate a further 8,000 houses. Yet writing as the Member of Parliament for an Oxfordshire constituency now adversely threatened – in the eyes of many of his more prosperous constituents – by the development of part of the 'rash of urban villages which developers seek to impose on green field sites', he takes a very different view, emotively referring to the countryside as being 'torn up and torn apart'. You will recall that we mentioned the proposed development of Stone Bassett in Oxfordshire in chapter 4.

To some this change of attitude on the part of Heseltine may smack of political expediency; to others it may indicate that the 'greening' of the Conservative Party, to which we referred in the Preface, had been of greater significance for some of its members. Certainly, few issues have caused as much internal disagreement in three successive Conservative administrations than those raised when the Draft Circular *Development Involving Agricultural Land*, was laid before Parliament in February 1987. But these differences in approach also surely throw into sharp focus the fact that when we consider what the countryside is and what we want it for and want for it,

11th March, 1988

Dear Mr. Ridley,

I am writing to you about the grave and growing anxiety in the
South of England about the self-evident ravages of the
countryside that the pace of development there is causing.

I understand full well the pressures to which you are subject.
Having held your post for four years I know of the conflicting
claims of economic growth, with its attendant requirements for
infrastructure and housing on the one hand, and the protection of
the environment on the other. You have an invidious task.

As a former Secretary of State for the Environment and as a
Member of Parliament representing a prosperous and beautiful
South Eastern constituency, I urge you to throw the weight of
your Department on the side of environmental restraint.

Wherever I drive in Southern England today, the place is being
torn up and torn apart. The pull of Europe and the completion of
the Channel Tunnel, both of which I strongly support, intensify
the pressure.

And, as you know, we now face a new threat posed by the rash of
urban villages which developers seek to impose on green-field
sites against the planning policies of national and local
government - and against the wishes of local communities - with
you forced to fight a field by field rearguard action through the
appeal system at taxpayers' and ratepayers' expense. There is no
solace in the argument that those urban intrusions will channel
inevitable development in a more co-ordinated way that would
otherwise proliferate along the urban fringe or by speculative
infill of every village. The truth is that we shall get both,
with an explosion of congestion in the areas concerned and a
burgeoning cost on the public purse - which will have to cope
with the consequent demand for educational and other services.
Indeed already porta-cabin classrooms testify to the gap between
demand and public provision.

Have you not the powers under planning legislation to indicate by
circular that you will not countenance such large scale intrusive
development? And can you not insure, or take powers to insure,
that when developers indulge in these extravagant endeavours then
the cost of resisting them will be met by the developers and not
the ratepayers?

You will detect in this cri de coeur the urgent need to extend
the urban programme of the Government to reclaim desolate and
derelict land in the run-down Northern and Midland towns and
cities to which development should be attracted. It must be a
priority today to protect the South by opening up opportunities
elsewhere.

I intend to make this letter public.

Yours sincerely,

Eileen Strathraver

Dictated by
The Rt. Hon. Michael Heseltine, M.P.
and signed in his absence.

The Rt. Hon. Nicholas Ridley, M.P.
Secretary of State for the Environment
Department of the Environment.

Figure 8.1 An open
letter to Nicholas Ridley.
Michael Heseltine,
former Secretary of
State for the
Environment in the first
of Mrs Thatcher's terms
of office, sets out his
own agenda for the
future of the
countryside.
Source: M. Heseltine

Figure 8.2 The urbanization of Oxfordshire? An aerial view of the site, next to the M40, for the proposed Stone Bassett development. Proposals like it in the South East, lie at the heart of Michael Heseltine's arguments against these new forms of development in the countryside.
Source: Consortium Developments

there are both national and local considerations which may at times either sit uneasily together or may indeed be at odds with each other. This is equally true if we first consider and then try to fit together in a coherent way, all the sector-based policies discussed in this book. There are many instances of this problem which we can cite.

In agriculture for instance, the need nationally has been identified by the present government, if not by its EC partners, as one which requires a considerable reduction in the over-production of key price-supported commodities and a stop put on the ever-escalating costs of the Common Agricultural Policy (CAP). Yet at the local level a more pervasive require-ment may be the maintenance of farm incomes, if not the sustenance of those local agricultural workers who have traditionally provided the life-blood of rural communities and economies.

As for the rural environment, the national need is to maintain and conserve areas of ecological importance and amenity landscapes whilst the need locally is to find ways in which these areas can contribute to the local economy.

How we can come to terms with this dichotomy is surely one clear test of a meaningful policy, or set of policies, for the countryside. This is a subject to which we will return later in considering prescriptive approaches to rural policy – those which specific interest groups would wish to thrust upon government as alternative if not, in their view, better policies for the countryside than its own.

But before that we need to remind ourselves just what those current government policies are, and whether they seem likely to meet its own avowed objectives especially when considered in relation to the financial resources available for their implementation.

Policies and their assessment

Without detailing again all that has been said in the preceding chapters, we must broadly be aware by now that the government's ALURE package consists of a largely fragmented collection of enactments and proposals aimed at enabling some form of readjustment to occur in a countryside that has been largely predicated on subsidized agricultural production, to one in which a much wider range of activities will prevail. You will recall that we discussed the specific measures to curb agricultural over-production in chapter 2 and proposals for expanding the productive and amenity roles of forestry in chapter 3. Inside a more flexible land-use planning framework, ultimately to undergo further reform as chapter 4 suggests, these activities are to be adjudicated and largely supported by market forces, as opposed to on-going government financial aid, although initially they may often be assisted by government grants and the newly-combined resources of CoSIRA and the Development Commission (now the Rural Development Commission) working inside their Rural Development Programmes. All of this is designed to stimulate a diversification of the rural economy, a potential that we assessed in chapter 5.

Within this climate of economic stimulation new measures are also proposed in the ALURE package and elsewhere for rural leisure and conservation. As we noted in chapter 6 the development of countryside recreation is to become much more promotional than hitherto has been the case and the number of conservation designations is to become much more comprehensive, particularly through the expansion in the number of Environmentally Sensitive areas (ESAs).

As if to provide a commentary on the government's package of rural measures first announced in the House of Commons in February of 1987, the Countryside Policy Review Panel, set up by the Countryside Commission in the same year, came to a very different set of conclusions about what was required to cope with the needs of an agricultural industry that was intended by government to lose just over one million hectares of productive capacity – all this in order to curtail the output of produce for which markets could not be found. That we should take notice of the Review Panel's findings is not just a comment on its distinguished and broadly-based membership, but that in reaching its conclusions it took into account comments from every national organization interested in farming and the environment – including the National Farmers' Union and the Country Landowners' Association. As one reputable journal put it 'the report is a definitive statement on how the government should proceed'.

As to the Review Panel's comments themselves, it reckoned that grants for diversifying farm enterprise needed to be higher by a factor of thirteen. Compared with the government's ideas of around £25 million it suggested a figure of £320 million. Instead of a government proposal of £7 million allocated to cover the cost of additional ESAs (over and above the originally designated six), the Review Panel reckoned that £30 million was nearer the mark to cope with between 30 and 40 sites. By the start of 1988, the government had doubled the number to twelve sites. As for diversification into woodland, the government believed a total of £13 million a year was needed to raise the total area over three years by around 33,000 hectares. The Review Panel felt that new woodlands should reach over 200,000 hectares in the period at a cost of £40 million a year. Whilst the £5 million planned by the government to cover farm diversification and marketing support seemed adequate, the Review Panel reckoned an *additional* £5 million was needed to aid the conversion of redundant buildings to new uses.

Since the initial House of Commons statement, the government has found it prudent to amend some of these figures. In November of 1987 the cash available in 1988–89 for alternative forms of land use, including the Farm Woodland Scheme, was increased by £10 million. For 1989–90 it also proposes to give hectarage grants to farmers undertaking extensification schemes, that is, growing less on the same area of land, as well as for schemes of setting land aside altogether, to the tune of £16 million. The funding of ESAs is also to be increased.

But these new totals still remain marginal compared with the Countryside Policy Review Panel's suggestions and when set against the £1,800 million that will be spent on agricultural intervention in Britain in 1988, as we noted in chapters 2 and 5. In addition to these financial sums, the government, in the judgement of the Review Panel, would also appear to have miscalculated the likely total reduction required of *land* in production since it reckons that just over 2 million hectares would have to go by the year 2000, twice the government forecast. Research workers at Wye College in the University of London support a figure of 3 million hectares, although other academic experts have gone for an even higher one of approximately 4 million hectares by that year. Our figure 2.7 in chapter 2 illustrates a set-aside of 3 million hectares.

Sins of omission

If a failure to come to terms properly with the changed circumstances of the countryside's key industry represents of itself a deficiency in the policy of the government, then another must be its lack of a coherent and comprehensive approach to broader rural issues. This is notably demonstrated by its disregard for what to many seem important aspects of social justice in rural areas.

Let us take housing, for example. In 1986–87, inner London council housing costs were met 25 per cent by central government, whilst the councils (mostly Labour) put in about 30 per cent from the rate fund. For the English shire district councils (mostly Conservative) both figures were a little under 2.5 per cent. Thus most of the public subsidy for housing (even on a 'per head' of population basis) would appear to be going to the urban under-privileged such as poor Londoners, rather than the deprived of the country-side. Moreover, although in late 1987 the government produced a White Paper and a Bill on housing of a relatively radical kind, which were aimed at eroding the powers of local authority housing departments, its approach still seems firmly centred on the housing problems of the big cities.

The particular difficulties of rural areas in terms of the housing market derive largely from the fact that wages are often low for those working in the countryside and yet the demand for more attractive rural properties is in many parts strong. This is a result of the desire of many of those who work in urban areas to live in a more tranquil and desirably rural environment. The 'right to buy' legislation of 1980, a strong plank in the platform of the new Conservatism, compounded this problem since it encouraged council tenants to purchase their own homes. Proportionately more did so in villages (well over 20 per cent) than in towns, partly because village estates are smaller and more attractive, and partly because the houses represent a particularly sound investment. Purchasers might qualify for a 40 or 50 per cent discount on market values, but the market values themselves are likely to be out of reach of sons and daughters of the village.

Clearly, recent government policies would appear in the eyes of most commentators to do nothing additionally to assist rural needs particularly in respect of this housing dimension, any more than the earlier Circulars *Land For Housing* (15/84) and *Developments Involving Agricultural Land* (16/87) did, with their pious sentiments regarding improved opportunities for rural housing. As a result, Rural Voice has clearly expressed the view that the Housing Bill of late 1987 should be modified to include a requirement that

Figure 8.3 'Sale of the Century?' Rural council sales provide short-term capital gains for the buyers, but a long-term loss of cheaper housing stock in the countryside. *Source*: Grenville Sheringham

the Housing Corporation earmarks a larger share of its overall budget for small scale housing schemes in the countryside, a sentiment echoed by the Countryside Policy Review Panel as we saw in chapter 1. This is now limited to a mere 3 per cent, although in England alone between 10 per cent and 20 per cent of the population live in rural areas, depending, as we argued in the first chapter of this book, on how you define them. Rural Voice is also inviting the Rural Development Commission to increase its contributions towards subsidizing the cost of expensive building sites.

Most important of all, Rural Voice demands a speedy review of the current relationship between housing and planning policies. Environment ministers have rejected the attempts of the Lake District to try to ensure, through the use of planning legislation, the availability of affordable housing to those members of its indigenous population who cannot afford prices often elevated by the demand for second homes in the area. As Rural Voice says, 'if government were questioning the means rather than the ends, it is important that it should come up with some other ways of ensuring such provision. Otherwise the drift of the less affluent young into the towns – priced out of village life – must continue posing an ever growing threat to the very fabric of rural communities.' It is not then surprising that a recent report of the rural issues group of the London and South East Regional Planning Conference stated that for the region 'the *foremost* need is to seek new ways to provide low cost housing since the situation is critical to the future of farm development and rural enterprise and the maintenance of the intrinsic value of the countryside'.

Meantime, efforts to counter such problems remain in the hands of small local initiatives, often prompted by the actions of Community Trusts. In one south-eastern county, Oxfordshire, the cheapest village houses are now over £60,000. The Stonesfield Community Trust, one of the earliest of a number of new attempts to strengthen community life through the creation of such umbrella organizations, is to have built high quality, low cost, starter homes. The key to achieving cheap housing here though, has been the gift of the land, currently fetching £1.2 million per hectare in the area, a factor which is unlikely to be replicated very often elsewhere.

However, a lack of council housing or housing at a price which can be afforded by the less well-off sectors of the rural community, is but one of many forms of deprivation in the countryside, as two surveys published in 1987 and based on data collected from a number of contrasting areas of the English countryside have made clear. These have highlighted the problems of rural deprivation amongst low income groups in general, but especially the pensioners and the families of manual workers, all usually without cars and dependent on inadequate and highly expensive public transport to get around.

Figure 8.4 Rural housing contrasts. Gracious living and housing deprivation increasingly co-exist as a result of a free market approach to rural housing.
Source: ACRE and Grenville Sheringham

Such problems of mobility are largely explained by the fact that government investment in rural transport remains low. Indeed a modest government grant of £1 million a year to a transport development fund run by the Rural Development Commission to encourage new ideas in rural transport, seems in keeping with its level of concern. A special bus grant for rural areas running at £16 million in 1988 is, perhaps predictably, due to fall to £10 million in 1989. This contrasts with £700 million of public money going into passenger rail services and £750 million into subsidies for other, largely town-based public transport.

Because of an inability to get around, both of these 1987 surveys have recognized that deprived groups inevitably have to rely heavily on the village shop, which has prices so much higher than the town-based multiples to which they have only limited access. For the old and the sick a five-mile round trip to the nearest doctors' surgery, by no means unusual, can be a further problem. Indeed, more recent evidence than that available from the surveys indicates an increasing problem resulting from the financial crises being faced by health authorities.

In counties such as Cornwall, Gloucestershire and Shropshire there has been a loss of many local maternity and geriatric services, cottage hospitals and in last-mentioned county, the hospital car service on which many of the less priviledged depended, to have access to health-care facilities. The innovative air ambulance service in Cornwall, so necessary in a county with a small scattered population, a terrain which is difficult to traverse quickly and with a heavy influx of holiday-makers in summer, is now having to depend increasingly on voluntary funding.

The general attitude of the government towards rural areas as far as health matters are concerned is, however, best summarized in the Primary Health Care White Paper *Promoting Better Health*, published in November 1987. Here the emphasis is on cost-cutting through the centralization of services, the effects of which must increase travel time and place higher financial burdens on patients. Whilst the impact of such policies must hit those who live in the rural areas hardest as we have seen, the White Paper fails to address such problems at all, consigning all references to rural areas to a marginal note in a chapter on the inner city!

Figure 8.5 The village shop in peril. This Dorset Post Office and store closed in 1987, despite a growing local population. Affluent newcomers offered too little custom!
Source: ACRE

As to the new influx of high tech industries to rural areas, this has brought little comfort to the jobless since these firms have taken with them their own labour forces. Of the opportunities which remain, as we saw in chapter 5, the skills which exist in traditional rural employment are not such as to be easily transferred to these new kinds of activities. Indeed, with educational opportunities and standards well below the average in the countryside, the possibilities for many rural school leavers in such enterprises is not great either.

With regard to the re-use of redundant farm buildings, an important aspect of the government's new policy for the countryside, these are already proving attractive although limited by planning consent to such uses as restaurants and craft workshops, these being seen as 'acceptable' alternative activities. As one of the organizers of the surveys of rural deprivation mentioned above put it, 'If you try to re-use buildings for car repairs and maintenance, you are in trouble. Yet the car repair man is modern rural Britain's blacksmith.'

Restructuring rural policies

Any attempts to come to terms with the social and economic problems evident in the countryside and discussed above, raise a number of key issues. These range from the kind of administrative arrangements that are likely to meet the needs of rural areas best, to how local economies can be developed to meet the needs and aspirations of local people and the ways in which the rural economy can be developed so that environmental quality is maintained or improved. The types of institutions and structures required to achieve a better integrated approach to the countryside and the level at which integration should occur are also vital issues of concern.

Such matters have been under consideration in a number of recent studies aimed at offering fresh approaches to the countryside. We have already made several references to the Countryside Policy Review Panel's views in this respect, including, for example, its advocacy of Rural Development Strategies as a tool for integrated rural management, as we saw in chapter 1, and for integrated recreation policies and networks referred to in chapter 6. But whilst the Review Panel's report makes such demands, it has been left to other studies to propose the means by which integration may be achieved. Some of these suggest the regrouping of local agencies and advisory services. Others, including Rural Voice, whilst praising the government for encouraging cooperation between the Ministry of Agriculture, Fisheries and Food (MAFF) and the Department of the Environment (DoE) as a 'move in the right direction', advocate the setting up of a Cabinet sub-committee with the task of tailoring important government decisions to the specific needs of rural areas. Most studies, however, after reviewing the wide range of statutory bodies at present involved in formulating and implementing rural policies and in the giving of advice, advocate a radical restructuring of administrative arrangements for the countryside.

The recent detailed submissions made to the government by the Association of County Councils in their report *Agriculture and the Countryside* and by the Royal Institution of Chartered Surveyors in their paper *Managing the Countryside – the Policy Framework* are examples of just such an approach. Certainly the conclusions reached by these demand an integrated solution, if not the creation of a Department of Rural Affairs, a line also taken by the Country Landowners' Association for over a decade as we saw in chapter 7.

But there are a considerable number of difficulties in bringing about such schemes and although in theory the ideal solution to the achievement of a more integrated approach at a central government level might be such a

Department, evidence suggests that this does not automatically follow if the European experience is anything to go by. In The Netherlands, for example, a single ministry, the Ministry of Agriculture looks after agriculture, forestry, conservation, recreation and regional development. It is rather like MAFF and parts of the DoE rolled into one. However, experience there suggests that more integration at national level can lead to rather more separation of individual interests at local level. Indeed, in The Netherlands there is little integration locally between agricultural, forestry or conservation. We must not forget, too, that the Forestry Commission is fully integrated into MAFF in Britain, but, again, there is little integration of the two sectors locally.

Some attempts by the government to reorganize British agencies have been in evidence. The loose-knit cooperation that now seems to exist between MAFF and the DoE is one example, whilst the proposed new role for district councils in land-use planning and the disappearance of the preparation of structure plans as a county council function, mentioned in chapter 4, offers another which could lead to a greater decision-making role for central government – a somewhat perverse development for an administration committed to deregulation!

However, such changes are far removed from the notion of a drastic restructuring of the kind discussed above which could prove expensive in terms of time and money and may be considerably resisted by interest groups within the Civil Service who wish for the maintenance of the *status quo* if previous experience of such matters is anything to go by.

An integrated approach – national and community initiatives

Since other European Community (EC) countries, as well as Britain, have their share of rural problems, the Community itself has been looking for better methods of tackling them. A recently formed Directorate of the European Commission was devised specifically to co-ordinate and integrate the spending of the main source of Commission funds, the 'structures' part of the agricultural fund, the social fund and the regional development fund. To begin with the Commission's concern in rural development involved structural measures such as the Less Favoured Areas Directive.

However, more recently, the Commission has set up several experimental development programmes, ranging from agricultural development programmes (ADP), through integrated operations (IO), to the most recently formed integrated Mediterranean programme (IMP). These, however, remain largely agriculturally based, even if it is their purpose to fund a wide range of projects across the economic spectrum in many of the more disadvantaged parts of Europe. Moreover, they do not necessarily embrace the key rural problems in any given area and in some instances have been applicable only to quite narrow problems at specific locations. Local people are also rarely involved with these schemes which seek to impose integrated solutions from EC level, rather than create a climate for them at the grass roots level.

Although a limited number of IO schemes have been set up, they have not always been successful. In the Western Isles of Scotland the rural development programme was perceived as a threat to highly-valued habitats and a vigorous debate ensued between the development and conservation lobbies. With regard to the IMP programme, cost has severely constrained its effective implementation with most of the effort so far restricted to Corsica.

As for Britain, some concern for a more integrated approach, albeit a very modest one, has appeared at government level. As we have already seen MAFF has widened its remit for the countryside, particularly under the 1986

Figure 8.6 Old bottles and new wine. Barns and other redundant farm buildings of character are being converted for such uses as antiques shops which tend to be more acceptable to planners than developments which might be of more direct benefit to the local community.
Source: CoSIRA

Figure 8.7 Developing the remoter parts of Europe. The Highlands of Scotland were one of the first parts of Europe to be designated a Less Favoured Area, by virtue of the danger of population losses. Since then Integrated Rural Development 'experiments' have been funded by the European Commission in the Western Isles.
Source: N. Curry

Agriculture Act, with support not only available for agriculture, but for tourism and the craft industries in Less Favoured Areas, for methods of farming which are environmentally sensitive in the ESAs and for farm woodland and other diversification schemes aimed at a reduction in the output of farm produce in surplus. The 1986 Agriculture Act also gives it more general recreation and conservation responsibilities. Such a notable departure from the support of agricultural objectives alone has also been underpinned by the ALURE package in which, as we have seen, a number of other departments are involved in supporting a much wider slice of the rural economy.

Perhaps more interesting than most in terms of government attempts to promote integration has been the legislation setting up the Norfolk Broads as a National Park in all but name. This had been achieved by the passage of a Bill through Parliament which became law in late 1987. The problems here have not so much been those of inadequate development as those deriving from the multi-resource use of a relatively small tract of countryside with great scenic appeal, ecological interest and recreational potential.

Although a Broads Authority has been set up in the late 1970s and vested with the task of trying to accommodate if not adjudicate between such diverse interests as agriculture, water conservation and supply, navigation and fisheries and recreation and tourism, it was notably failing. This was because control of these sectoral activities remained in hands other than those of the Authority. Its role remained advisory whilst the very fabric of the environment of the Broads seemed in terminal decline.

Figure 8.8 A National Park in all but name? Sustained pressure over a long period to make the Norfolk Broads a National Park has lead instead to the development of a unique management structure for the area.
Source: Countryside Commission

The run-off of fertilisers from the local fields and the disposal of treated sewage into the river system by the water authority was causing the eutrophication of the Broads themselves; farmers were ploughing the grazing meadows, a key feature of the traditional landscape; holidaymakers' motor boats were damaging the banks of the waterways by their speed and by the sheer weight of their numbers. However, the passing of the Broads Act has sought to remedy such a situation by giving the Authority statutory control over these activities and by the implementation of a properly coordinated policy in an attempt to conserve the area for the benefit of all its users.

Local integrating initiatives

Such an Act of Parliament, along with other top down policies promulgated by central government and the European Commission, has largely evolved as a response to the activities of interest groups, a matter we have touched on in several chapters in this volume and discussed in some detail in chapter 5. But although they represent some kind of positive redress of the balance of competing demands on the countryside, as well as an opportunity to spread tax revenues across a broader spectrum of Community concerns, they do not automatically stimulate bottom up local developments. Indeed, early examples of local integrating initiatives were, by their pre-dating many central government projects, bound to be less a function of government policy and more a result of effective cooperation at the local level between a range of public and voluntary sectoral organizations.

As we saw in chapter 5, these initiatives were specifically designed to meet the aspirations of the local rural community. This is true of the Staffordshire Moorlands Project set up in the late 1970s by a number of agencies. MAFF which had long been criticized for wearing agricultural blinkers was an active partner in this scheme, the outcome of which was a project which embraced tourist development, farm labour sharing and new local employment opportunities funded by CoSIRA. Although ADAS as a participant inevitably focused on the agricultural sector, it even then recognized that the future of the rural economy lay as much with non-farming development as with the farm sector.

In the early 1980s another experiment was established which partly overlapped the Staffordshire Moorlands Project. Here the Peak National Park was the lead agency in the Peak Park Project which involved two small villages, Longnor and Monyash. Also supported by all the other local rural agencies, and part-funded through the European Commission's Integrated Operations Programme, it resulted in the setting up of a new alternative grant aid system. This partly replaced, but also largely augmented, the existing sources of social and economic support. Because of its success in providing a better environment and improving employment prospects it was extended to a third parish in 1985.

Other similar rural development projects have been set up in the Cotswolds, the Lleyn peninsula in north Wales, in the Teifi valley south-east of Aberystwyth, in the Waveney district of East Anglia and at Leominster in Herefordshire. However, that the emphasis in all of these has been on social matters and job creation (often working in association with the Manpower Services Commission) rather than just the environmental dimension, owes much to the lead taken in setting them up by Rural Community Councils. Nevertheless, other schemes using the fully-integrated model as represented by the Peak Park Project do exist, as in Cumbria, or are planned.

The Sussex Rural Community Council, in association with the Countryside Commission, the Rural Development Commission and East and West Sussex County Councils, is, for example, starting and financially supporting a completely integrated scheme in the High Weald Area of Outstanding Natural Beauty. This will involve, to start with, the two parishes of Forest Row and Dane Hill. But the project is fortunate in having as its leader a person with experience of integrated rural development in other EC countries (including France, Italy and The Netherlands) and in Austria where such ideas are well developed. Indeed, in all successful schemes of integrated rural development, apart from a shared ethos and a significant input of personnel from the collaborating agencies, they have in common a

Figure 8.9 A tale of two villages. Longnor (left) and Monyash in the Peak District National Park have been the subject of the best known integrated rural development experiment of the 1980s.
Source: Countryside Commission

considerable dependence on a front man or woman of energy and ability. It is these animateurs who have the key task of acquiring and maintaining the confidence of the recipient communities. We discussed their potential in chapter 5, where we examined the role they might have to play in 'signposting' information and sources of grants.

Local integration – French style!

The locally-integrated schemes for rural development that we have discussed so far remain, however, *ad hoc* and unstructured. Although they do utilize policies and agencies put in place through central government and as a result of EC initiatives, there is no formalized link between the two main levels that could, if pursued, properly lead to a more effective approach and offer a better future for our countryside.

In France, however the situation is rather different. There a framework does exist which, although supported mainly by the department responsible for agriculture, is also endorsed by other relevant departments of central government and is geared towards the development of local programmes of action which are initiated at that level. These not only harness the benefits of national and EC policies, but then combine them with local resources to meet local needs. In order to activate this approach, groups of communes, each roughly equivalent to a village and its parish, are encouraged to come together to plan and follow through initiatives which receive central funding.

Arrangements of this kind depend partly for their success on a much higher level of power vested in the communes than would be the case in similar administrative units in Britain; in France these lowest rungs in the three tier system of local government have responsibility for development control and land-use planning.

As part of the initiation process, groups of participating communes, assisted by the higher regional and departmental echelons of local government, are first involved in an extensive data collection and collation exercise relating to the key problems of the area concerned. This leads on to a programme of local action which may turn out to be based on a broad mixture of national policies and local development initiatives, or may favour either end of such a spectrum. Final agreement to proceed is sanctioned in a

Figure 8.10 Community developments in France. Community-built facilities in a commune in the Pas de Calais house a school, teachers accommodation, commune office and village hall.
Source: ACRE

formal agreement between the group of communes and regional and central government.

That such a system exists at all owes much to the decentralization legislation in France of the early 1980s which came into effect in 1983. In essence this enabled a situation to arise where leading roles in promoting initiatives of an action-based kind could be undertaken by communes leaving the formerly more powerful regional and departmental governments with roles of support and coordination, roles which are perhaps important but not fundamental.

But can Britain take it?

It may seem a curious moment to move quickly from French achievements in the sphere of schemes of local rural activism which build on central policies, to a consideration of the basic political ethos of the present British government. However, it may just be that this French model not only has much to recommend it in terms of achieving a true balance between national and local needs, the point with which we started this chapter, but is in sympathy with the current political climate in Britain, a matter which we touched upon briefly in the previous chapter and chapter 4.

After all, a recent analysis of ten years of the new Conservatism by a group of political scientists, emphasized its philosophy of replacing central government 'as the major agent of allocation of goods, services, benefits, burdens and costs' and 'rolling back the frontiers of the state . . . substituting market forces wherever possible for state action. Where it has not been possible to replace state provision', it concludes, 'considerable encouragement has been given by this government to enable market forces to play a role'.

As if finally to knock on the head the notion of achieving any form of integrated or highly structured rural policy administered from Whitehall, it emphasizes the present government's belief that administrations in principle should not coordinate information or policies centrally. The government, believes it can and should only provide the framework of law and stability within which individuals or groups make their own economic decisions and control their own destinies. 'Planning is inefficient and free markets do a better job of promoting growth, relieving poverty (at least in absolute terms) and coordinating individual decisions, than any amount of planning.' All this seems entirely consistent with a spirit of opportunism and innovation at the local level merely working inside a framework of support provided by central government; indeed, just those elements of the French model.

But if the different British administrative structures make it difficult to use the model directly, certain basic underlying principles nevertheless have their appeal in the context of this country. For example, the principle of the block funding of key rural initiative programmes present in the French model already has a parallel in our Urban Aid Programme and ought to be readily transferable. Thus whilst budgetary control could be maintained by government with respect to capital sums, local decisions could be made about spending programmes. This would contrast with the open-ended commitments to funding by which so many of the agricultural support policies have recently been characterized.

However, a number of other sources might still be tapped although the range of grants available, for example, inside the ALURE package, might well be more flexibly administered. Interestingly enough, the French experience shows that a competitive spirit has developed amongst the separate agencies offering financial support and this has been of positive advantage in enabling specific projects to take wing. Similarly, the key role of the animateur is already apparent in his or her ability to secure funding and schedule development projects.

Another principle that might be translated to the UK is that of the freedom to choose to engage in development or not. In Britain local authorities have a legal obligation to participate in some aspects of planning even if it is only to produce local and structure plans. If the French model supposes no compulsion to engage in local development initiatives, the government merely providing an enabling framework or structure, experience shows that those who do so engage can reap considerable rewards.

Once again this must emphasize the importance of the animateur in the initiation and following through of schemes of action. Clearly, the few local initiatives in Britian mentioned above have also depended on someone playing an equivalent role. If these animateurs are at present few and far between, perhaps a parallel also exists in many of the Rural Community Council 'patchworkers'.

Finally, the French model presupposes an applicability of procedures for development proposals to any part of the rural environment; central government policies and local support can be provided wherever community action is required and there are central government advisers available to assist in their setting up. This contrasts with Britain. Here, as we have seen, support for agriculture of an environmentally acceptable kind is currently primarily directed to certain areas such as Less Favoured Areas and ESAs; management agreements are used almost exclusively in SSSIs or National Parks and so on. The important fact here is that the location is a primary determinant of whether support is granted at all and, if so, from which source.

Nevertheless, it remains true in reality that much of the wider countryside, if not all of it, requires some level or another of support. The translation of some of the philosophy of this French model to the UK might make this

Figure 8.11 The human catalyst. 'Patchworkers' have a vital role to play in the success of rural community development schemes.
Source: N. Curry

possible, especially since a radical restructuring of existing administrative frameworks is not called for.

These adjustments to the British system with reference to comparative initiatives represent only one way forward. There are other advocates of alternative approaches as we have already seen. But when the Association of County Councils put its case to government for a unitary approach to the resolution of the problems of the countryside in which the needs of agriculture, the environment and rural society were to be brought together in a coherent way, its response was not enthusiastic. As the Association reported 'it favoured a case by case approach rather than an attempt to devise a comprehensive framework'.

Given such a political climate the French model seems a far more plausible way forward; ideas of integrated planning with an emphasis on a top down structure are definitely not on the political agenda in this country for the present or for the near future.

Such speculation about this model, for that is what it is, could seem idle if the government fails to address it in the foreseeable future. However, that would do its discussion in this chapter less than justice, for its very examination in relation to Britain has underlined a number of valuable points, not least the way in which political ideologies impinge upon rural policymaking and the importance of the perspective we addressed at the beginning of this chapter – the dichotomy of national and local rural needs.

Figure 8.12 Farming with imagination. The government is now actively seeking to encourage the production of conservation grade produce, but this organic farm in Wiltshire has been going for 15 years.
Source: R. and O. Ellis

If we are to address fully the theme of a future for our countryside it is equally valid to consider that for the immediate future its destiny might be merely the outcome of the policies that we have seen put in place since early 1987. If our horizons are thus lowered, what is it we might look for?

Down to earth

Let us start with the farming scene. Those who stay in agriculture and are not eased out by a remorseless downward pressure on prices for supported commodities, or are not even coaxed out by government incentives, may be

those who have so far least come to terms with environmental considerations; in other words, those that have high capital inputs and are free with the chemicals. As one distinguished rural correspondent recently so neatly wrote 'it takes imagination to be an organic farmer, while science will continue to make it easier to be lazily productive'. Thus it will be the less efficient, 'the dog and stick' farmers that will be forced out. These are essentially the 'middle band' farmers that we mentioned in chapter 2.

Clearly, if the present Minister of Agriculture is to be believed, and all the evidence of pressures now being exerted on Brussels by EC member countries suggests it, we are not in a situation where in a few years there will be no support for farmers. Instead of a bottomless pit, he wants to see the Community provide a safety net, its provision being in the interests of both producers and consumers. If that proves a comfort to farmers, will conservationists be equally pleased with a possible outcome from set-aside policies?

It is said that these could result in some parts of farms being run in accordance with high standards of environmental protection with the rest producing crops with even less concern for the environment than at present. But having announced that the scheme would fulfil environmental as well as economic objectives, the Minister of Agriculture feels that such fears are groundless, especially if the details of set-aside are properly worked out in the first instance. Whether the Secretary of State for the Environment also has it in mind that Rural Conservation Areas, which we mentioned in chapter 4, might additionally promote such a balance, we do not know.

Certainly for the moment there is little evidence to sustain a view one way or the other, although the Countryside Commission, having rejected the idea of Rural Conservation Areas, continues to call for the plans for set-aside as so far revealed to be redesigned so as to offer very much more by way of conservation and recreational benefits, as well as reducing food production. Whether some areas of land are worked even harder to counter falling prices, or 'the dog and stick' farmers go to the wall, what seems certain is a further decline in the numbers employed on the land. Equally inevitable in the longer run will be the added complication of a need for fewer and fewer hectares to produce the agricultural commodities we do want as bio-technology and genetic engineering inexorably increase yields.

If set-aside schemes do not work either to cut surpluses or improve the environment, then there will be pressure to further expand the ESA programme (always assuming that those already designated function properly and the fears expressed about them in the previous chapter are quite without foundation!).

Certainly, a reduction in price support for sheep and beef or greater competition from lower cost lowland producers seeking to diversify could lead to a situation in hill farming where demands for ESA status might seem the only means of staying in business, apart from forestry. This status could be seen not only as a social support mechanism for such farmers, but as a device for the maintenance of an environment which might otherwise increasingly take on an appearance of neglect if agriculture withdrew from the hard uplands. At present the government finds ESAs too expensive to be considered as a wide-spread phenomena, but ultimately we could be drawn into another National Health Service situation where additional money is intermittently dragged from the government purse under the relentless pressure of media attention. The government then might find it easier to adopt a somewhat freer attitude to that part of the rural domain left outside of ESAs, apart from that otherwise specially protected either because of the quality of the land itself or the landscape.

Where such areas are closer to centres of economic growth, as in the South

East, footloose entrepreneurs could have a field day. Certainly, the demographic 'hole in the doughnut' syndrome whereby the central city functions fly to the periphery of the urban area and beyond could be a planners' nightmare if the cash starved 'dog and stick' farmers decide to cut and run. That is unless the planners are persuaded by some positively expressed view of government into not just a docile acceptance of the inevitable, but to see the situation as a challenge to design a semi-urbanized countryside redolent of the best of Milton Keynes – our 'Spread City' of chapter 4. However, this is again mere speculation since there is no evidence of government thinking along such lines.

Meantime the problem of rural deprivation in terms of housing, health, education and transport, accumulate with little in the government's philosophy to persuade it to do much about them, although it remains under sustained pressure from the Rural Development Commission. This body is at least saying that if the new breed of rural artisans that the government expects to appear in our countryside as a result of its non-agricultural diversification policy do materialize, then housing provision will need to become a major priority.

With their arrival the post-war rural wheel would have turned full-circle. The majority report of the war-time Scott Committee whose ideas were married to the view that 'every agricultural acre counts' and enshrined in so many Acts and policy documents over a period of more than forty years,

Figure 8.13 'Spread City'. A vision of the rural future – a semi-urbanized countryside stretching from Cambridge to the south coast and from Oxford to Dover.
Source: Sunday Times

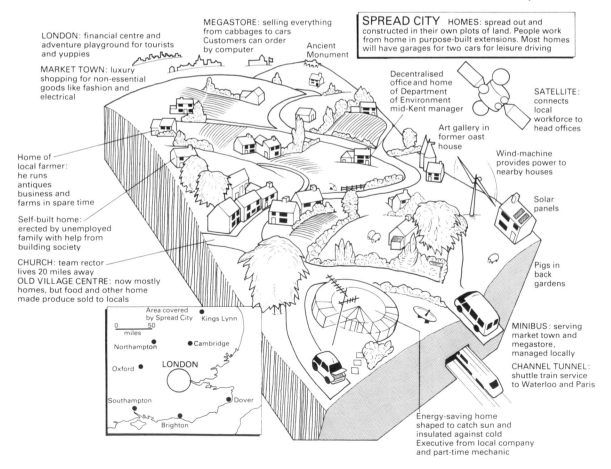

would be finally dead and buried. Professor Dennison's minority view expressed in that report and which envisaged a countryside no longer intrinsically dependent on agriculture, but with a broader economic base supported by a wide variety of rural industries, would seem, at last, to have come to pass.

Figure 8.14 'A Future For Our Countryside?'
Source: *Earth Mirth*

Further reading

From a changing countryside to a changing policy response

A good start to understanding the background to recent policy developments will be to have a look at *The Changing Countryside* and *The Countryside Handbook*, both published by Croom Helm (London, 1985). As predecessors of this volume, they explain and analyse many of the policies for the countryside up to the early 1980s and provide a brief account of much of the legislation, organizations and so on, that have influenced these changes. *The Changing Countryside* is also useful because it takes you deep into the history of key rural issues, from that of increasing agricultural productivity, to the containment of urbanization, through the conservation of the wild to the sustaining of rural communities.

Further reading for the rest of this chapter really concerns the policy documents themselves. The Ministry of Agriculture, Fisheries and Food's *Farming UK* (HMSO, 1987) provides a useful summary of most of the policies that we discuss relating to agriculture. This is an annual review of farming, though, so you can keep up-to-date by looking at the subsequent editions. They may have slightly different names from year to year. For forestry you might look at the National Audit Office's *Forestry in Great Britain* (HMSO, 1986) which presents a view on the economic consequences of forestry, and the policy consequences that flow from them.

To read about land-use planning and rural diversification policies, have a glance at the Department of the Environment's *Rural Enterprise and Development* (HMSO, 1987). This, too, has its annual equivalents, so do keep an eye open for them. For recreation, the Countryside Commission's *Enjoying the Countryside* (1987) provides a useful summary of policies and proposals. It also has an accompanying video which you might like to glance at. The Department of the Environment's *Conservation and Development* (HMSO, 1986) summarizes government conservation initiatives over the previous ten years.

A general review of rural policy proposals is presented in the Countryside Policy Review Panel's *New Opportunities for the Countryside* (Countryside Commission, CCP 224, Cheltenham, 1987), which tends to be a bit more wide-ranging than official government statements.

Countering the food problem: too little action for too much food?

There is a large and often repetitive literature about the CAP and its reform, much of it put out by the European Commission. One official document which is worth reading is *Perspectives for the Common Agricultural Policy* (European Commission, Brussels, 1985). This sets out the Commission's own informative analysis of the different policy options and includes a useful review of socio-structural policy. Among the many academic commentaries on the CAP, Brian Hill's *The Common Agricultural Policy* (Methuen, London, 1984) is a well written and accessible analysis of how the policy works. *The Food and Farm Policies of the European Communities* by S. Harris, A. Swinbank and G. Wilkinson (Wiley, Chichester, 1983) provides a more detailed, if slightly dated, sector by sector analysis.

For a cogent and very readable study of the political economy of the Common Agricultural Policy you should try *New Limits on European Agriculture* by F. Duchène, E. Szczepanik and W. Legg (Croom Helm, London, 1985). The book paints a faintly depressing picture of unresolved contradictions in agricultural policy and points to the insuperable limits to a continued expansion of the industry. For a spatial perspective and a very clear analysis of some of the unintended distributional effects of the CAP, the rural geographer Ian Bowler has written *Agriculture under the Common Agricultural Policy: a geography* (Manchester University Press, Manchester, 1986). A number of crisp polemics have been written about agricultural policy in recent years. Richard Howarth's breezy critique *Farming for Farmers?* (Institute of Economic Affairs, London, 1985) is a particularly good example of the genre.

On the quickening debate surrounding the CAP and the environment you should read David Baldock's elegant essay *The CAP price policy and the environment: an exploratory essay*, published in *Can the Common Agricultural Policy Fit the Environment?: The Environmental Implications of Future EEC Farm Price Policies* edited by David Baldock and David Conder (Council for the Protection of Rural England, London and Institute for European Environmental Policy, London, 1986).

Expanding forestry and woodland: barking up the wrong tree?

The historical background to present-day forestry is admirably covered in N. D. G. James's book *A History of English Forestry* (Blackwell, Oxford, 1981) which spans nine centuries from Roman times. Oliver Rackham sets woodlands in their wider countryside context in *The History of the Countryside* (Dent, London, 1986).

The first fifty years of the Forestry Commission are chronicled in G. B. Ryle's *Forest Service* (David and Charles, Newton Abbot, 1969) whilst the Forestry Commission's *Annual Reports* contain detailed information on its activities. Valuable comments on the Commission's economic performance are contained in two HMSO (London) publications: the 1986 report by the National Audit Office, which we mentioned in the first part of 'Further Reading' and the twelfth report from the House of Commons Committee of Public Accounts, *Forestry Commission: Review of Objectives and Achievements* (April, 1987).

The mid-1980s have seen a flurry of reports and statements on forestry from public bodies and voluntary organizations. These include the Nature Conservancy Council's *Nature Conservation and Afforestation in Britain* (Nature Conservancy Council, Peterborough, 1986) which analyses past practice and looks for a better accommodation between forestry and nature conservation, and, in the voluntary sector, the British Association of Nature Conservationists' *Forests of Britain* (Packard, Chichester, 1987). This attacks upland afforestation on environmental grounds and argues for new patterns of control and incentives. The Countryside Commission's *Forestry in the Countryside* (CCP245, Cheltenham, 1987) attempts to identify a balanced approach towards multiple forestry goals and examines ways of implementing national forestry policy.

T. Denne, M. J. D. Brown and J. A. Abel in *Forestry: Britain's Growing Resource* (UK Centre for Economic and Environmental Development, London, 1986) look at recent forestry policy, examine options for the future and usefully add a European dimension. Advocates of privatization will be interested in Robert Miller's controversial free-market approach in the Hobart Paper *State Forestry for the Axe* (Institute for Economic Affairs, London, 1981). Another thought-provoking contribution to the debate has been provided by P. J. Stewart in *Growing Against the Grain: UK Forestry Policy* published by the Council for the Protection of Rural England (London, 1987).

A more flexible system for planning and development?

The most recent national study of containment, Green Belts and countryside planning is *Green Belts: Conflict Mediation in the Urban Fringe* by Martin Elson (Heinemann, London, 1986). John Herington's *The Outer City* (Harper and Row, London, 1984) also discusses development trends and public policy responses in the pressured countryside.

Issues of land ownership, land values and land management are fully treated in *Development and the Landowner: An Analysis of the British Experience* by Robin Goodchild and Richard Munton (Allen and Unwin, London, 1985). A critique of the underlying values of our system of land-use planning is provided in *Whatever Happened to Planning?* by Peter Ambrose (Methuen, London, 1986).

A more polemical account of the impact of lobbying groups in countryside planning and the roles and actions of MPs is given in M. Davies' *The Politics of Pressure: the Art of Lobbying* (BBC Publications, London, 1985). A lively case study of central/local conflicts over planning in the green fields of the M4 motorway corridor in central Berkshire is offered by John Short and his colleagues in *Housebuilding, Planning and Community Action: The Production and Negotiation of the Built Environment* (Routledge and Kegan Paul, London, 1986).

A useful synthesis of the special pressures on the urban fringe is found in the Countryside Policy Review Panel's report of 1987 mentioned in the first part of 'Further Reading'. The problems of the land-use planning system and some proposals for change are convincingly set out in *Town and Country Planning* (Nuffield Foundation, London, 1986). A good up-to-date account giving a flavour of the problems facing local authorities in the countryside is to be found in *Report of the Rural Issues Group* (London and South East Regional Planning Conference, RPC 923, London, 1987). The government's draft proposals for changes to the land-use planning system are set out in *The Future of Development Plans: A Consultation Paper* (Department of the Environment, London, 1986).

The stances of the two sides of the Tory party on the balance to be struck between conservation and development are best seen in *This Green and Pleasant Land: A New Strategy for Planning* (the Conservative Political Centre, London, 1987) which offers the view of the 'wets'; and *An Environment for Growth* (Adam Smith Institute, London, 1987) which is considerably 'drier' in its approach.

Diversifying the rural economy: the real stimulus to rural wealth?

The diversification of the rural economy and the resultant employment consequences are not a subject that has been widely written about in depth in spite of plenty of journalistic coverage. However, there can be no better introduction to attitudes to the countryside than a combination of Raymond Williams, *The Country and the City* (Chatto and Windus, London, 1975) and Howard Newby's *Green and Pleasant Land?* (Wildwood House, London, 1985, 2nd edition). For attitudes to work in the post-industrial society *The Future of Work* by Charles Handy (Blackwell, Oxford, 1984) provides an illuminating and readable introduction.

The general framework of national employment change is looked at in two books. Manufacturing employment change is explored by S. Fothergill and G. Gudgin in *Unequal Growth; Urban and Regional Employment in the United Kingdom* (Heinemann, 1982), whilst J. Goddard and A. Champion (eds) have also looked at employment change from a national perspective in *The Urban and Regional Transformation of Britain* (Methuen, London, 1983).

The most significant study of rural employment change is Ian Hodge and Martin Whitby's *Rural Employment: Trends, Options, Choices* (Methuen, London, 1981), although some might criticize its rather pessimistic outlook. Other information on changing trends in rural employment is to be found in chapters in books. Two chapters in *Rural Britain: A Social Geography* by D. R. Phillips and A. M. Williams (Blackwell, Oxford, 1984) provide background information, whilst at a more theoretical level and from an international perspective K. Hoggart and H. Buller's *Rural Development: A Geographical Perspective* (Croom Helm, London, 1987) is useful.

There are two valuable additional sources that are published regularly. *The Countryside Planning Yearbook* (Geo Books, Norwich) edited by Andrew Gilg, provides an annual review of academic literature and some good review articles notable amongst which is G. Williams' contribution to the 1984 *Yearbook*. In 1987 this volume took on an international dimension and should continue to be useful although perhaps less comprehensive in its new guise than before. The other source is the annual conference proceedings of the Rural Economy and Society Study Group. A number of papers in *Deprivation and Welfare in Rural Areas* (Geo Books, Norwich, 1986) by P. Lowe *et al* explore certain aspects of rural employment.

On-farm diversification is covered by Bill Slee in *Alternative Farm Enterprises* (Farming Press, Ipswich, 1987) and in S. P. Carruthers (ed.) *Alternative Enterprises for Agriculture in the UK* (Centre for Agricultural Strategy, Reading, 1986). The consequences of part-time farming are comprehensively discussed in Ruth Gasson's *The Economics of Part-time Farming* (Longman, London, 1988).

As you will by now be well aware, government departments and quangos produce annual reports. Expensive but useful here is the annual report of the Rural Development Commission which contains a reasonable, if mildly self-congratulatory, summary of the current set of policies and their operation. We have already mentioned the Ministry of Agriculture, Fisheries and Food's recent glossy publication *Farming UK* in the first part of 'Further Reading'. In the context of rural diversification it provides some good background information on the policy winds blowing through Whitehall.

Increased rural leisure: recreation for all?

If you would like to know more about people's leisure patterns, the Countryside Commission's National Surveys of Countryside Recreation are summarized in *National Countryside Recreation Survey 1984* (Countryside Commission, CCP 201, Cheltenham, 1985). More general leisure trends are comprehensively reviewed in a book by Chris Gratton and Peter Taylor entitled *Leisure in Britain* (Leisure Publications Ltd., Letchworth, 1987).

This covers countryside recreation and tourism, as well as other leisure characteristics.

Social policies for rural leisure are reviewed in a paper by Nigel Curry and Alison Comley, entitled *Who Enjoys the Countryside?* (University of Strathclyde, Department of Urban and Regional Planning, Occasional Paper No. 9, 1986). Recreation Transport issues are comprehensively discussed in the Manchester University study that we referred to in the chapter. It is written up in David Groome and Chris Tarrant's paper *Countryside Recreation, Achieving Access for All?* in the Countryside Planning Year-book (Geo Publications, Norwich, 1985).

Land-use policies are discussed in Alan Patmore's book *Recreation and Resources* (Blackwell, Oxford, 1983). A more exhaustive treatment of them is presented in a research report by the Centre for Leisure Research entitled *Access to the Countryside for Recreation and Sport* (CCP 217, Countryside Commission, Cheltenham, 1986). The legal aspects of land-use policies are given a thorough treatment by Tim Bonyhady in *Law of the Countryside: the Rights of the Public* (Professional Books, Oxford, 1987). This is particularly useful on common land and rights of way.

For implementation and management, *Interpretation in Visitor Centres* by the Dartington Research Institute (Countryside Commission, CCP 115, Cheltenham, 1978) provides one of the few comprehensive analyses of people's reactions to interpretation facilities. The best analysis of recreation pricing can be found in *Recreation Management and Pricing* by Tony Bovaird, Mike Tricker and Robbie Stoakes (Gower, Aldershot, 1984).

Finally, the contemporary policy initiatives for rural leisure of the Countryside Commission are presented in their booklets *Policies for Enjoying the Countryside* and *Enjoying the Countryside: Priorities for Action*, both produced in September 1987.

Conservation: more than appeasement?

The problems surrounding the implementation of the 1981 Wildlife and Countryside Act have been highlighted in two main texts. There is W. Adams' *Nature's Place: Conservation Sites and Countryside Change* (Allen and Unwin, London, 1986), as well as *Countryside Conflicts: Politics of Farming, Forestry and Conservation* by Philip Lowe *et al* (Gower, London, 1986). An examination of some of the broader political implications of the shifting terms of the debate during the years immediately after the passage of the Act is provided in 'Agriculture and Conservation in Britain: a community policy under siege' in the book edited by G. Cox *et al*, *Agriculture: People and Policies* (Allen and Unwin, London, 1986).

A discussion of the Halvergate incident is contained in 'Halvergate: the Politics of Change' by T. O'Riordan in the *Countryside Planning Yearbook* (Geo Books, Norwich, 1985). *Broads Grazing Marshes Conservation Scheme 1985–1988* (Countryside Commission, CCD 20, Cheltenham, 1988) describes the operations and achievements of the scheme designed to prevent the further destruction of the traditional Broadland grazing marshes. The evaluation and monitoring of the Environmentally Sensitive Area programme, which has its beginnings in the Broads conservation scheme, will be the subject of a Ministry of Agriculture, Fisheries and Food report to Parliament in 1988. Meantime, a powerful argument for the extension of the ESA concept (albeit in a more sophisticated form) to cover all the National Parks appears in Ann and Malcolm McEwen's latest book *Greenprints for the Countryside* (Allen and Unwin, London, 1987).

FWAG has been the subject of a lengthy study by G. Cox, P. Lowe and M. Winter, *The Voluntary Principle in Conservation – a study of the Farming and Wildlife Advisory Group* is to be published by Packard (Chichester) late in 1988.

One of the most astringent conservationist critics of agriculture in the pre-1981 period, Marion Shoard, has recently resumed the battle in *This Land is Our Land: The Struggle for Britain's Countryside* (Grafton Books, London, 1987). More measured tones are adopted by Norman Moore, a former chief advisor to the Nature Conservancy Council, in his book *The Bird of Time: The Science and Politics of Nature Conservation – a personal account* (Cambridge University Press, Cambridge, 1987).

New rural policies to the turn of the century: is the sum greater than the parts?

Of the literature mentioned in this chapter, the report of the Countryside Policy Review Panel, already referred to above, is of major interest with its well-argued discussion of integrated policies. *Agriculture and the Countryside*, a report prepared by the County Planning Officer's Society (1986) for the Association of County Councils, primarily advocates an integrated approach to the countryside through a Ministry of Rural Affairs, although it also makes seven other recommendations. This report is obtainable from 66a Eaton Square, Westminster, London SW1W 9BH. A rather more thinly argued case for integrated solutions, appears in a memorandum by the Royal Institution of Chartered Surveyors *Managing the Countryside – The Policy Framework* (1987) and is available from 12 Great George Street, Parliament Square, London SW1P 3AD.

As for 'bottom up' approaches to integrated rural development Ken Parker's *A Tale of Two Villages* on behalf of the Project Steering Group, Bakewell, the Peak Park Joint Planning Board (1984) tells the story of Monyash and Longnor, whilst *Rural Viewpoint*, the bi-monthly magazine published by ACRE (available from Stable Yard, Fairford Park, Fairford, Gloucestershire, GL7 4JQ) regularly reports on similar innovations as well as being atuned to the broader social problems of the countryside. On the previous point, of the two reports on rural deprivation referred to in the chapter, the main one, prepared by Brian McLaughlin for the Department of the Environment and entitled *Rural Britain in the 1980s: Rural Deprivation Study*, remains unpublished. However a *Summary of Findings* is now available free of charge from Mrs J. Rudman, the Department of the Environment, Room N19/14, 2 Marsham Street, London SW1P 3EB. The second of these documents *The People, the Land and the Church* (1987), edited by Richard Lewis and Andrew Talbot-Ponsonby comes from the Hereford Dioscesan Board of Finance, The Palace, Hereford HR4 9BL.

Housing problems in the countryside have also been highlighted in *Village Homes for Village People*, available from the National Agricultural Centre Rural Trust, 35 Belgrave Square, London SW1X 8QN. This recommends that the Housing Corporation should especially recognize rural housing needs and that local authorities be allowed to spend more to fund charitable housing associations.

The French model for rural development is discussed in detail in *La Décentralisation; La Dynamique du Développement Local* by P. Coulmin (Syros: ADELS, Paris, 1986) if your French is up to it! The political background of Britain in the later 1980s, which in the last analysis must be a key factor in determining future rural policies is discussed in *Developments in British Politics* by Henry Drucker (et. al.) (Macmillan, London, 1988).

Finally, we cannot end without drawing your attention to the many useful journals that can sustain and nurture our concern for rural policies. Apart from the already-mentioned *Rural Viewpoint*, the bi-monthly *Countryside Commission News* (published from John Dower House, Crescent Place, Cheltenham, Gloucestershire, GL50 3AR) can keep you well informed about Commission initiatives, usually pointing you in the direction of its more detailed 'in-house' reports.

A number of organizations concerned with the countryside also publish journals primarily aimed at their membership but of interest to a wider readership. The monthly *Country Landowner* from the Country Land-owners' Association (16 Belgrave Square, London SW1X 8PQ) is surprisingly broad in its concerns, even reporting such matters as rural housing problems, and the Council for the Protection of Rural England (4 Hobart Place, London SW12W 0HY) has its quarterly *Countryside Campaigner* which is highly informative about campaigning issues which the organization is earnestly pursuing. *Rural Wales* is the equivalent publication from the Council for the Protection of Rural Wales (Ty Gwyn, 31 High Street, Welshpool, Powys SY21 7JP). But many other organizations also have useful regular journals details of which you can check out in *The Countryside Handbook* already mentioned at the beginning of 'Further Reading'.

Turning to conservation in particular, *ECOS*, the journal of the British Association of Nature Conservationists, is an excellent source of comment, topical articles, news items and reviews in this area. Also quite strong on conservation issues is the new commercial bi-monthly *Environment Now* although it spreads its coverage to other areas of rural concern, as well as beyond these shores.

Newspapers, and two in particular, are useful in keeping up-to-date in rural policy issues in general. *The Observer* from time to time treats such matters in depth of which Robert Chesshyre's 'Cry the Beloved Countryside' of 2 August 1987 is an outstanding example. Richard North continues to do exemplary work in his frequent analyses of rural issues in *The Independent*.

Index